BUST?

Also By Robert Peston

WTF?
How Do We Fix This Mess?
Who Runs Britain?
Brown's Britain

ROBERT PESTON
& KISHAN KORIA

SAVING THE ECONOMY,
DEMOCRACY AND OUR SANITY

First published in Great Britain in 2023 by Hodder & Stoughton
An Hachette UK company

1

Copyright © Robert Peston and Kishan Koria 2023

The right of Robert Peston and Kishan Koria to be identified as
the Authors of the Work has been asserted by them in accordance
with the Copyright, Designs and Patents Act 1988.

All rights reserved. No part of this publication may be reproduced, stored in a retrieval system, or transmitted, in any form or by any means without the prior written permission of the publisher, nor be otherwise circulated in any form of binding or cover other than that in which it is published and without a similar condition being imposed on the subsequent purchaser.

A CIP catalogue record for this title is available from the British Library

Hardback ISBN 9781399700757
Trade Paperback ISBN 9781399700764
ebook ISBN 9781399700788

Typeset in Plantin Light by
Palimpsest Book Production Ltd, Falkirk, Stirlingshire

Printed and bound in Great Britain by Clays Ltd, Elcograf S.p.A.

Hodder & Stoughton policy is to use papers that are natural, renewable and recyclable products and made from wood grown in sustainable forests. The logging and manufacturing processes are expected to conform to the environmental regulations of the country of origin.

Hodder & Stoughton Ltd
Carmelite House
50 Victoria Embankment
London EC4Y 0DZ

www.hodder.co.uk

For our families,
who came here as immigrants
and, like us, believe in this country

CONTENTS

1	The Point of Being King	1
2	Bust, in a Nutshell	14
3	It's Intelligence, Jim	33
4	Lessons in Catastrophe	64
5	A Devalued Bank of England and Treasury	104
6	Margaret Thatcher is Dead	135
7	Growing Pains	164
8	Bye-Bye Boomer	197
9	Learning How to Lose	222
10	Artificial Hope and Real Hope	245
	Acknowledgements	269
	Index	271

CHAPTER 1
THE POINT OF BEING KING

My ninety-two-year-old mum is fortunate to be able to get by on my late dad's university pension. She lives in a part of North London that is a community of every income and ethnicity. Until she broke her hip and going out on her own became riskier, the local shopkeepers in a street of kebab takeaways and all-night grocers looked out for her. A glue in this community is the nearby Arsenal football stadium, the Emirates. Everyone wants the Gunners to win. There's pride in a team whose diversity is a mirror to London. When I walk home from the match, store owners and parents pushing prams will ask, 'How did we do?' These symbols of our identity matter, where they unite but don't exclude. We'll welcome anyone as a Gooner, even those born in Tottenham.

My sister Juliet and I chat about Mum most days. She and my brother Ed are her unpaid carers. If it weren't for them, Mum would struggle to cope. The risks of independent living would be too great. Jul and Ed are part of the UK's vast informal social service. There are millions of Juls and Eds, who selflessly support dependents old and young. Usually, these carers are acting on the idea that a healthy family is the corollary of a healthy society, a balance of obligations, duties and rights. As a nation we usually take this unremunerated service for granted. But it is a kind of small miracle. Given the crumbling state of public services, imagine the mess we'd be in if vulnerable people were abandoned by those who love them. These are the people whom the state should reward with knighthoods, damehoods and gongs, not the civil servants and business leaders whose careers have already given them money, status and job satisfaction. Their conviction of an entitlement to this kind of public recognition, in the form of a title, is pervasive and insidious. This country's values often feel like *fin de siècle* Paris and Moscow, as chronicled by Maupassant and Tolstoy. Gordon Brown once regaled me with how he had been assailed by the wife of a distinguished politician because the order of knighthood he had been awarded was the 'wrong kind' and she knew he was deserving of a superior one. I didn't fully appreciate there was a hierarchy of knighthoods. It never occurred to me that even all 'sirs' aren't equal. Which would be comic if it didn't reflect a kind of moral bankruptcy in the privileged class. More than any other modern prime minister, Boris Johnson recognised the magnificent bankruptcy of the honours system, showering them on friends, supporters, political party donors and hosts of lavish weekends away. He gave peerages to a brother and aides barely thirty, and he wanted to give a knighthood to his father. If we saw this kind of thing happening in another part of Europe, we'd describe it as a racket. This is Britain though.

In the bartering for influence, power and public recognition, Adam Smith's invisible hand is wonky. The market for the prizes of status and recognition is inefficient and irrational. There is, still, grotesque systemic bias. The children of assimilating upwardly mobile Ashkenazi Jews from London's East End, I acquired the remorseless ambition and thick-skinned confidence. My sister didn't. But she had the brains. Therapy may help me properly understand the origins of this difference. It hasn't so far. Juliet chose to be a chef and was brilliantly creative, until she chose to leave the kitchen. I chose my more visible career in journalism and television. Absent a less sex-and-gender-skewed power structure, her voice in the public space would be louder than mine. I've experienced a fair amount of antisemitism in my life, but mostly I thank my lucky stars to have been born a white male Baby Boomer. It could have been a good deal worse. And when Kish tells me about the racism that almost became mundane for his parents and grandparents, my sense of privilege is reinforced. So when Juliet rings me, it's frequently to make sure that I am seeing the world as intelligently and sensitively as she does.

Mostly, Jul laments how the country seems to have become less fair and kind than what we expected of it when we were growing up in the 1970s. The day after the coronation of Charles III, she was fuming that three volunteers in Soho, so-called 'Night Stars' authorised by Westminster Council to patrol streets at night and protect the safety of women, had been arrested a couple of nights earlier. They were carrying rape alarms. And rather than accept their explanation that these are the tools of their philanthropic trade, the police assumed they were anti-monarchy saboteurs, intent on setting off the sirens to panic the horses during the gilded parade. It was as though the agents of the state have lost trust that anyone can be public-spirited.

The previous September, the days around Queen Elizabeth's death had a similar sense of unreality and madness to a death in the family. That was true for those of us who work in the Palace of Westminster and in the media. On the morning of

8 September, there were rumours around Parliament that the Queen had suffered some kind of seizure. I spoke to a number of privy councillors, the senior politicians who are supposed to be the monarch's advisers on the deployment of the royal prerogative. They were flustered, anxious for news, not unlike characters in a Shakespeare play. It is on days like this that the constitutional reality – that she is head of state, endowed with the formal power to make or break elected governments – reveals itself in all its momentous significance. There was concern in Parliament greater than I have experienced when general elections are called. This was, of course, partly because of fondness for the only monarch many of us have known. But it was also because the constitution was about to face a test stiffer even than the transfer of power just days earlier from a disgraced prime minister, Boris Johnson, to one chosen by just 81,326 Tory members, Liz Truss. There weren't riots on the streets that the country would have a new political leader in the absence of a general election. The legitimacy of Truss replacing Johnson rested on the perceived legitimacy of a Queen who anointed the new prime minister in the ceremony known as 'kissing hands', which was at Balmoral Castle on 6 September. And Queen Elizabeth's authority rested on more than seventy years of service to the United Kingdom as monarch. It could not be taken for granted that her son, the soon-to-be Charles III, would have anything like the same authority. And if he didn't, then what?

The Queen died shortly after 3pm. In a very British way, we at ITV – and I assume throughout the rest of the media – knew, but obeyed the convention that we wouldn't reveal until there was an official statement 'from the Palace'. Hilaire Belloc's ditty still captures something of the relationship between established power and the press:

Bust?

You cannot hope to bribe or twist
Thank God! the English journalist:
But seeing what the man will do
Unbribed, there is no reason to.

Deference has been under siege for years. But it is not vanquished. I was instructed by my ITV bosses to install myself in the road in front of the door to 10 Downing Street, to deliver my reflections on the implications for the new prime minister of the death of the Queen. There was a practical problem though. I had come into work in a suit and tie that were far too cheerful, and my more sober ones were at the cleaners. So I frantically tried to obtain a dark suit and a black tie. Then, when I got to Downing Street, wearing the black tie, the programme editor panicked. If they came to me live, before the formal announcement of the demise, but after we'd already reported she was gravely ill, viewers would see the black tie and fear the worst. So there was then a frantic search for a tie that was sombre but not uniformly black. It was as though the whole constitutional confidence trick – the conceit that there is a rational basis to the role of the monarchy – would be imperilled if any of us made a false move.

My attitude to the British monarchy is very much like my attitude to my Jewishness or religion more widely. I am a sceptical agnostic about rabbi and king, rather than a militant atheist or principled anti-monarchist. I don't find it easy to celebrate a system in which a billionaire white landowner should by dint of birth be head of the state I am proud to call home. And the spectacle of our elected representatives fawning to the new King is uncomfortable. I worry about what this manifestation of the advantages of inherited power and wealth must do to the motivation of children growing up in poverty in jobless households. If I look at other democratic models, an electoral system for the head of state is not necessarily or always associated with more stable and less corrupt systems of democratic government. At the time of writing, America's democracy – in the wake of the storming of the Capitol on 6 January 2021 – looks more vulnerable than ours. But simply by being a woman in the man's world of the latter half of the twentieth century, and by winning trust and respect from almost all she encountered, the story of Elizabeth II is a much better look for the United Kingdom than Charles III can ever hope to paint. Most of this is not his fault. He didn't choose his biological sex. And some, though not all, of the intense scrutiny he's received all his life, even from the lackadaisical journalists of Belloc's rhyme, has largely revealed someone with human rather than unforgivable frailties. But it means neither he nor we can assume his constitutional position is unassailable.

On coronation day itself, 6 May 2023, I was surprised therefore that Juliet bought a commemorative coronation teddy bear for Mum, with Charles III's crest on its paw and sporting a purple bow tie, from the Sainsbury's supermarket across the road. It turns out this was a triumph. Because Mum – though never a sentimentalist or avowed royalist – loves the bear and carries him everywhere. The actual King can only dream of inspiring such devotion. What struck Juliet about the supermarket where she acquired Charles Bear was the emptiness of the shelves, coronation kitsch aside, and she asked the store manager what was going on. He told her that the incidence of theft, even of basic stuff like washing powder, had increased so much that the shop had a policy of keeping the shelves sparse till the security guard arrived later in the day. It was a commercial response to pilfering that is on the rise.

No one shoplifts detergent and cleaning products for kicks. People were stealing because they were desperate, caught in the vice of stagnating incomes and soaring prices of the basics of life, from electricity to bread and milk. There has been a starkly differential impact of this inflation on businesses that have pricing power and people on low incomes with no negotiating power. Just days earlier, Sainsbury's had announced 'underlying' annual profits of £690m, at the top end of expectations, from a 5% rise in sales. This was slightly down on the previous year, but almost a fifth higher than before the Covid-19 pandemic. In the year to 4 March 2023, the remuneration of the chief executive, Simon Roberts, was £4.95m, an inflation-busting rise of 26%, during what was for most of the country an acute cost-of-living squeeze. It was 229 times the pay of a typical or median Sainsbury's employee. Over at bigger Tesco, the chief executive, Ken Murphy, earned £4.44m, a fall of 7%. Murphy earns 197 times the median pay at his massive supermarket company. The very British backdrop to this weekend of gilded coronation pageantry was inequality that is more extreme than in most of Europe.

The coronation was a moment to take stock of who we are. It didn't happen. Our leaders indulged in predictable platitudes that fought against the evidence of our eyes and ears. The prime minister, Rishi Sunak, put out a statement that there would be 'a vivid demonstration of the modern character of our country', and 'a cherished ritual through which a new era is born'. The leader of the opposition, Sir Keir Starmer, wrote in the *Daily Telegraph* that: 'this weekend, an estimated 300 million people will tune in from all around the world to see Britain crown a new King. As they watch those pictures, they will see a national celebration. This is our country at its finest – enjoying our history and our traditions, while raising a glass to the future.' Both said what they assumed King and country would expect to hear from them. Original thought was in short supply.

I asked Starmer in an ITV interview what exactly it was about the anointing of a billionaire landowner as head of state and head of the established church that appealed to someone like him, whose party was created to promote equality and to battle unfair privilege. 'It's the sense of our country coming together,' he said. 'It's a symbol of hope for our country . . . People desperately need that hope and that sense of togetherness. And I agree with them. I think most people are fed up with the politics of wedge, wedge, wedge, divide, divide, divide, let's split people up.'

As part presumably of his preference for balm over 'wedge, wedge, wedge', Starmer could not bring himself to express outrage after the police arrested those who wanted to exercise their right to protest against the coronation and others wrongly identified as a threat to the processing cavalry. All he could bring himself to say was that the police had to learn from their mistakes, as new powers they've been given to limit disruption from protests 'bed in'. He's right that culture wars and identity politics are eroding our common purpose, the fabric of community. But fealty to the monarch is perhaps the oldest extant form of identity politics in the longest-running of culture wars, that of 'my country is better than yours'. We must be able to be more creative when healing the fractures in our nation than simply to wave a Union flag as a gold carriage glides past.

Labour shares with Sunak and the Tories an appreciation

of the putative benefits of the UK's constitutional monarchy, convinced it glues us together, gives us a sense of belonging and displays a marketable image to the world. These are the justifications you'll also get if you ask one of the various new artificial-intelligence chatbots to make the case for a constitutional monarchy such as Britain's. They are not profound.

Across the world's democratic nations, other constitutional models are available, but with everything feeling so fragile, our leaders are terrified even of having the debate about whether we should move to an elected head of state. Nor do they seemingly question the monarch's role as head of an established church, even though it represents an identification between state and church that is surely incongruous in a country of so many faiths and none. Their refusal even to think about whether inherited privilege is an appropriate constitutional model is problematic, when we have just blown £100m on a coronation binge of processions and dressing up, during an epidemic of mental ill health and acute stress among those who can't pay for food, rent and heating. Meticulously rehearsed processions and glittering swords distract from a country that is failing in many different ways. It's circuses instead of bread. At the end of coronation day, my ITV colleague Paul Brand wrote, 'it was a moment of competence and celebration, the likes of which Britain has seldom enjoyed in recent troubled years.' He was right. The complicated choreography of marching, fly-pasts and sword-bearing went off flawlessly, if dampened by overcast skies. But the idea – seemingly promoted by both Sunak and Starmer – that this kind of 'competence' is going to arrest our national decline is worrying. It can't.

Here is the nutshell. We are living through technological and economic shifts over which we have limited control.

There is a global race for leadership in industries that are being invented or totally reconstructed by artificial intelligence. The winning nations will be those that attract the brainiest people, the expert knowledge and the capital. Now think of the coronation as a £100m global advert or marketing exercise. Ask yourself if a single additional pound of investment in UK tech or life sciences venture would be triggered by televised pictures of a sixty-seven-year-old white male Old Etonian archbishop placing a jewel-encrusted crown on the head of a seventy-four-year-old white male alumnus of another fee-paying school, who'd first been anointed in 'holy oil'. What message is being sent when the new King's brother, Prince Andrew, kitted out in the robes of the highest chivalric order, the Garter, was given a front-row seat, after having paid £12m to settle a rape accusation (which he denied)? The chaotic negotiation of Brexit and the budget crisis of the autumn of 2022, when Liz Truss was prime minister, damaged the UK's reputation for legal and constitutional stability, and depressed investment flows to the UK. Why would advertising to billions of people that power and money in the UK is an accident of birth reverse that?

If you are in the business of trying to finance or create the new business opportunities created by artificial intelligence, how likely are you to deploy your money or know-how to a country whose people have just been asked to pledge in unison as follows: 'I swear that I will pay true allegiance to Your Majesty, and to your heirs and successors according to law'? There's a place for hocus-pocus and irrationality in life, but surely not as the brand image that any nation wishes to present to the world? The risk is that the UK is seen either as a national version of a Disney-style theme park or as a benighted theocracy. Nor is it likely to appeal to the talent in countries where British imperialism meant oppression and enslavement for their predecessors.

For a UK that's in trouble – too short of political inspiration, economically stagnant, unable to properly fund vital public services, befuddled by the self-righteous screams in a cultural civil war – the coronation was displacement activity. And perhaps the biggest cost of the monarchy is how *we* relate to it, because we use it to reinforce self-defeating British exceptionalism, our sense of being special. It's an opioid that traps us in the lazy thinking and complacency that stops us confronting and fixing our practical problems. We need to analyse dispassionately the challenges we face and why the belief in progress is in such short supply. We should not hide from our problems under an ermine robe.

Despondency is greatest among those with their whole lives ahead of them. A poll of 18-to-25-year-olds by Savanta for *Peston*, my ITV show, revealed that almost two-thirds of young people have little or no trust in politicians, and that 61% expected to be no better off than their parents during their working lives. Their pessimism is well grounded. Even though there was much greater unemployment when I was entering the workforce in the early 1980s, social mobility was greater, and general economic prospects were better too. Today's generation is the first since the Second World War to know nothing but a crash followed by stagnation.

If, by some miracle, you are not worried about the future, about how things aren't getting better and won't get better, if you have total confidence in our leaders, then this book isn't for you. But it matters why hope has vanished. We desperately need to restore it. The inheritance of and the prognosis for the UK and the West are disappointing. Growth and living standards have been stagnating for fifteen years.* There has been a relentless worsening of inequalities, stemming from structural changes to the economy that disproportionately reward the wealthy. There are threats to our livelihoods and even to our existence from climate change, from real and incipient pandemics, from the hostility of states whose values are different from ours, notably Russia and China, and from the technological revolutions of artificial intelligence and quantum computing. But it's not all danger. The opportunities from AI and from fixing climate change could also be transformative in a very positive way. To seize the benefits, we have to open our eyes to the dangers. Although historical comparisons can obscure as well as enlighten, the nearest analogous decade is the 1930s. That's because of the scale, nature and complexity of the challenges. It took a world war to purge the world of Nazism and fascism, and to initiate the slow and steady proliferation of

---

* On 1 September 2023, the Office for National Statistics revised up its estimate of the size of the British economy in the fourth quarter of 2021 by 1.8%. This revision does not alter the medium-term view of the economy, namely that since the 2007/8 financial crisis growth has been disappointingly low. It also tells us little about the UK's future prospects because it largely relates to the unpredictable stock-building behaviour of firms and the output of hospitals during the almost-unique two years of the global Covid-19 pandemic. The new higher GDP number is also irrelevant to the reality of the savage squeeze in living standards being endured today by British people. It may be reassuring to the prime minister Rishi Sunak that the UK's economic performance during the Covid-19 years of 2020-21, when as Chancellor he spent a record £400bn keeping us afloat, no longer seems materially worse than other competitor economies. But all this means is that much of the West, especially Germany and much of Europe, shares the UK's low-growth problem - though who knows what the final picture of relative performance will turn out to be, when other countries follow the UK's lead and revise their statistics for that challenging era?

market-based democracies, as Martin Wolf chronicles in *The Crisis of Democratic Capitalism*. Today the confidence and solidarity that underpin a sustainable democracy are daily denuded by the lies that proliferate on social and mass media. Don't assume our way of life is safe.

The United Kingdom may disintegrate if we don't urgently reinforce the economic ties that bind, and if voters were to decide that Westminster is irredeemably incompetent and opt instead for narrowly focussed nationalism in all four nations, including England. Culture wars that sow mistrust between ethnicities, genders and sexualities could yet become white hot. The foundations of the state, trust in democracy, are not immutable.

One of the most important ties that bind, the late Queen, has gone. It matters whether the new King can show leadership on the question of whether we can rehabilitate Britain as a place where doing your bit, working hard and playing by the rules will deliver a decent income and home. Is this even possible for a white man in late middle age who inherited his title, hundreds of millions of pounds in property and his role as head of state? On the Nixon-goes-to-China principle, that the biggest breakthroughs are sometimes delivered by the most unexpected people, he could surprise us. If he's not thinking hard about the responsibilities of a head of a state in an era of potential national fracture, we may be in difficulties, and he may be in even worse trouble.

CHAPTER 2
BUST, IN A NUTSHELL

We've known since at least the banking crisis of 2008 that the system isn't working as it should, for the West, for the United Kingdom, for you. The political explosions since have seemed important as they happened but have done nothing to reverse decline. In 2010, voters kicked out a Labour government that had been in office since 1997. In 2016 they voted to take the UK out of the European Union. In the 2017 general election, substantial numbers – 40% of those voting – backed the most left-wing Labour Party of modern times, whose leader Jeremy Corbyn was promising to nationalise on a scale not seen since the late 1940s. In the 2019 general election, a whopping 43.6% voted for a Tory Party whose leader, Johnson, promised to 'get Brexit done' and to 'level up' the country. All of these votes were a different version of the same song, one that demanded a new direction for the country. Voting for Corbyn and voting for Johnson were different choices, usually by different people, but they were both rejections of a status quo that was failing. According to analysis by the *Financial Times*:

> *In 2007, the average UK household was 8 per cent worse off than its peers in north-western Europe, but the deficit has since*

ballooned to a record 20 per cent. On present trends, the average Slovenian household will be better off than its British counterpart by 2024, and the average Polish family will move ahead before the end of the decade.*

The lack of earnings growth is acutest in the public sector. Only those in the private sector are now earning a little more in real terms than they were before the 2008 crash. One way of seeing the squeeze on the pay of doctors, nurses, teachers, police officers and the rest is to look at their position relative to private-sector pay. According to analysis by the Institute of Fiscal Studies, public-sector employees earned on average 19% more than private-sector employees in 2011, at the onset of George Osborne's public-sector squeeze, austerity. This is not altogether surprising. Public-sector workers are typically educated to a higher level and have more advanced skills than those in the private sector. Even so, last year, the gap between public-sector pay and private-sector pay had fallen to just 7%, the lowest margin between the two categories since 1993.† And if adjustment is made for the differential skills of public- and private-sector workers, there is now zero difference between the sectors, which is unique in modern times.

It's not just that the UK is falling fast down the international league table of living standards. Much of the fabric of the state is fraying. Statistics on how little crime is being solved by the police are – well – criminal. According to the Casey Review, the most important police force in the country,

---

* John Burn-Murdoch, 'Britain and the US are poor societies with some very rich people', *Financial Times*, 22 September 2022

† Boileau, O'Brien and Zaranko, 'IFS Green Budget 2022: Public spending, pay and pensions', The Institute for Fiscal Studies, October 2022

the Metropolitan Police, is institutionally sexist, racist and homophobic. There are 7.47m waiting for elective care from the NHS, up from 4.39m just before the pandemic. Some 22% of school students are persistently absent from school, which means they are missing more than 10% of lessons. And since Covid, almost half a million have disappeared from the workforce, many of them suffering from mental and physical ill health.

Our way of life is worse than it was fifteen years ago. Perhaps because there have been so many enormous shocks – banks collapsing, an illness that imprisoned us in our homes, the invasion of Ukraine – we are less angry with the government than perhaps we should be. In the case of Covid-19 and Putin, we see the disasters as exogenous, not the government's fault, however we may assess the subsequent management of these crises. We cut our leaders slack, knowing it's been harder than normal to fix the roof with flood water pouring through the front door. But among the many crises, Brexit was an active choice of so many leading politicians, not an accident that happened to them. And Brexit was always going to make us poorer. That was the basic economics of increasing the costs of trading with our most important market, the European single market. It's what I pointed out night after night on ITV as its political editor during the referendum campaign. Sunak of course knew this, even though he argued at the time that leaving the EU would unleash the dynamism of the UK. Johnson should have known this. So what has been surprising has been the slowness of successive Tory governments to develop a comprehensive industrial strategy to counterbalance the immediate downward pressure on prosperity. Voters' tolerance of substandard stewardship won't be inexhaustible.

Neither Kish nor I are fatalists. We don't believe decline is inevitable and irreversible. We are realists who believe progress is possible when illusions are abandoned. In this wealthy country, bursting with talent, despair is absurd and self-indulgent. Let's see Britain and the world as it is, and build from there.

One of the reasons we wanted to collaborate is because we have different experiences on which to draw. We are separated generationally and ethnically. I am a tail-end Baby Boomer and Kish is a Millennial. I am a secular Jew; Kish's family are Hindu. We're united by a love of a country that has provided us with so many opportunities to be the people we want to be. And we're also united by a perception that it's wrong – immoral – to ignore or minimise the trouble faced by the UK and much of the West.

When we talk about the country being bust, we don't just mean in the narrow financial sense of a government struggling under the weight of its debts and interest payments. The 'bust' is a more generalised breakdown. One manifestation is a kind of borderline personality disorder that touches millions of us and our governing class. On social media and in the mainstream press, we pick needless fights with each other over gender, race, sexuality, faith. And successive governments have engaged in cheap squabbles with our closest allies in the European Union, except when faced with the kind of unifying threat that is Vladimir Putin's Russia. We have a mutually destructive tendency to focus on our otherness, to push people away rather than embrace and celebrate difference. Periodically, we suffer from dark despair that we'll never amount to anything much again. And then we swing to equally irrational euphoria about being so much better than anyone else.

I am not starry-eyed about the past. I was a teenager in the messy 1970s: inflation that peaked at 25% in 1975, industrial unrest, power cuts, price controls, an IMF bailout, rubbish uncollected, bodies unburied. But it was also exciting and liberating, hopeful even. Bowie and T. Rex challenged gender stereotypes. Bowie's androgyny on the BBC's *Top of the Pops* was a gender-interrogating licence to paint nails and wear make-up. *Monty Python*'s surrealist comedy – about the Spanish Inquisition and the great philosophers playing football – made it cool to be a swot. The Sex Pistols rhymed 'anti-Christ' with 'anarchist', and the Clash celebrated the Sandinistas. Ian Dury set his working-class chronicles, about teenagers stealing top-shelf mags, his bedsit-living dad and Dickie, the Billericay lothario, to music in the great Victorian music-hall tradition. Bob Marley led a reverse cultural colonisation. Popular culture was dumbing up, not down.

My belief in a better Britain and a better world had both generational and ideological underpinnings. In my comprehensive school in North London's Crouch End, my classmates' parents came from India, East Africa, West Africa, the Caribbean, and the Turkish and Greek parts of the island of Cyprus. I was the only Jew in my class. The only time in the classroom I thought I experienced mild antisemitism – although perhaps I was being oversensitive – was when my English teacher teased me for not noticing that Michelangelo's David isn't circumcised. I loved that teacher, who was gay and out at a time when that took courage. There were only a tiny number of indigenous whites, and racism felt absurd as a belief system, impossible in our community. I saw and experienced racism outside of school. But I was confident it would be consigned to the dustbin of history. This turned out to be naive.

I had a similar confidence in progress with respect to sexual preference and identity. Homosexuality between consenting adults over the age of twenty-one was legalised in 1967. The toxic stigma attached to coming out had not been eliminated. I felt pride that my sister and a small number of her friends defied conventions to reveal they were lesbian. It was harder for boys to express a preference for other boys. No boy thought it was anything other than funny to call another boy a 'poof'. But despite the daily reminders of residual prejudice, I was confident the barriers to sexual self-determination were on their way out. History was on the side of progress.

The evidence was in front of our eyes that pretty much everything that mattered was getting better. When my mum and dad hired their first colour television, it was one of the most exciting days of my childhood. On that curved Baird screen, I thought I was seeing the whole world properly for the first time. Every year there was a new and life-changing technological leap: video recorders, an electronic ping-pong game, 'Pong', that plugged into the television, the Sony Walkman. Every new food was a door to greater cultural understanding: Chinese and Indian, hummus, a doner, melon and Parma ham.

I was living proof of upward social and economic mobility. My parents were second-generation immigrant stock, the children of tailors and pleaters, who didn't just become middle class but were insistent they had risen to the upper middle. I had no doubt that my own life chances would be even better than those of Mum and Dad, even though there was mass unemployment when I left university in the early 1980s. It was also obvious that my school friends, pretty much whatever their backgrounds, were likely to lead more successful and longer lives than the previous generation. This is not rose-tinted projection from forty-five years on. These were the conversations I had as an eighteen-year-old, and they turned out to be prophetic.

All the economic, political and cultural volatility ushered in huge change. 1979 was a fulcrum, a pivot point. To be clear, it didn't seem inevitable there would be an overhaul of the state by the most ideologically right-wing government in British history, that of Margaret Thatcher. As a cocksure teenager, I knew another less brutal path was available for the UK. Today I can see the point of Thatcher and of Thatcherism, something almost like historical necessity. Unlike the post-Brexit paralysis, she fomented an economic revolution through privatisation, liberalisation of the City, tax simplification and tax reductions that ultimately delivered rewards to many millions of people, although swathes of the country, and especially Scotland, will never forgive the way she wrecked their industries and livelihoods.

Even though she was a more divisive prime minister than pretty much any before or since – and even though there was no shortage of conflict on the streets, over the closure of coal mines, the smashing of print unions by Rupert Murdoch, her reform of local government taxation – there was no serious challenge to democracy, no sense that losers in elections would not take the rough with the smooth. We were, oddly, a more united nation, even in the very angry divisions over her and her policies.

Kish's formative years were much later and calmer, in the nineties and noughties, and in Kent rather than London. This was the Blair/Brown era, which felt stable for him, personally and politically. His grandparents had endured the stress of settling here as East African Indian migrants and his parents applied themselves to win a better life. It is a familiar story of living and working in corner shops, driving forklifts in factories and working in the NHS. Life for them was in London, Luton, South Wales and eventually the prosperous South East. Self-improvement was periodically punctuated by bricks through windows, the hostility of the

National Front and uneasiness about whether they'd ever truly belong. Kish's life was much easier; although he was the token member of an ethnic minority in the communities he inhabited when growing up, he was never uneasy about fitting in. He had the love and support of his mum, who chose to be at home in his younger years in an era where single-income families were still something that was possible for many. His preoccupations were the same as those of his school friends. The self-interrogation was whether Man United, Arsenal or the newly rich Chelsea would win the league, whether he would be able to have a new pay-as-you-go mobile phone and (more important, but less urgent) if he had any idea what he wanted to do with his life. For most of Kish's friends, the life choice took a while to reveal itself, but having the freedom to choose was an extraordinary privilege, which his parents and grandparents never had.

Britain was much more at ease with itself than today. Beckham, Britpop and the Spice Girls were the UK's advert to the world. They represented a new generation but were no threat to the status quo, co-opted as they willingly were into a monarchy-loving patriotism, always smiling, tolerant and inclusive. Global economic change – disinflation from a rapidly industrialising China, rampant consumerism – made our money go further, made us richer. The internet meant that a whole world of entertainment and distraction was increasingly at Kish's fingertips and was generally considered more interesting than the steady, dull progress of a three-term Labour government. Political and social revolution didn't feel a sensible option. The big stuff was under control. Kish was interested in politics, though, because – unlike many of his peers – he was a nerd and proud.

After the initial excitement around the election of Tony Blair in the 1997 landslide – his earliest political memory – only the war in Iraq really stood out as a moment of political significance that truly punctuated the consciousness of Kish's circle of comfortably off young people. I, my late wife Siân Busby and my then five-year-old son Max were among 1.5m on the streets of London protesting against the looming invasion that would be ordered by Bush and Blair. Our shouted voices were unheard by Blair, or indeed by the majority of ministers and MPs. At the time, a majority of British people were opposed to the invasion, unless or until there was a so-called second resolution at the United Nations that legitimised the military action – a resolution that never materialised. This was an occasion when our representative democracy failed but – like the 1980s – there was still a clear national consensus that our flawed system was superior to the alternatives.

It was the banking crisis that shook Kish's unspoken assumption that economic progress would be seamless. Until then, he and the majority of his friends took for granted that Britain was on the up, although in his rural and affluent South East neighbourhood Labour were never flavour of the month and rising anxiety about immigration saw a younger Nigel Farage come to prominence.

Britain's constitution – uncodified though not unwritten as the common view would insist – was remarkably robust throughout. Our patch-and-mend liberal democracy remained relatively free from challenge, even though millions of British people did not, for almost a generation in the eighties and nineties, secure their preferred government. Participation in elections did fall, however, from 76% in 1979 to a still-legitimising 71% in 1997. There were small numbers who supported extreme groups of the left and right. But debate was largely about how to strengthen rather than dismantle the foundations of the UK's democracy. In fact the one earthquake of the 1980s, the split in the Labour Party to yield the Social Democratic Party, was explicitly manufactured to give a new home to political moderates, those for whom revolutionary change was total anathema. The greatest achievement of the SDP was arguably to force the Labour Party subsequently to remake itself in the SDP's image. There might not have been Blair without it.

The British way was to complain, often very rudely, about the government and its ministers, but not to shake the political infrastructure that they rested on. There was a lot to complain about. After Margaret Thatcher's administration was elected in 1979, unemployment rose to a peak of almost 12% in 1984 – three times where it is today – and the numbers of children living in poverty increased from around 15% to 28% by 1990. For whole communities wrecked by the closure of mines and steel plants, these were wretched times. But across the world, democracy was on the march, because if anyone examined life in the big autocracies, they saw grinding poverty in much of China and bleak corruption in the Soviet Union. Democracy was conspicuously better than the alternatives. And somehow the Victorian, male and public-school-dominated British system spawned a state-school-educated woman, who trained to be a chemist, as the most influential prime minister of the second half of the twentieth century. Change was possible.

Thatcher was brutal in consigning heavy industries to the dustbin of history. She was simplistic in her conviction that the private sector always manages better than the public sector. She castrated trade unions and empowered corporate leaders to pay themselves however much they liked. But she engineered economic rehabilitation. I refused to see it at the time, but I am persuaded that her privatisations, her tax cuts and her labour market reforms helped to stem the UK's economic decline, though – and this is important – only because the country was part of Europe's common market and then helped to turn the common market into the European Union. It was precisely because a more competitive United Kingdom had access to the European single market – a huge seamless market that Thatcher and her officials played a leading role in designing – that the country enjoyed all those years of growing prosperity from 1992 to 2007. It is part of the current British tragedy, or perhaps comedy, that the influential politicians on the right wing of the Tory Party, who seek a return to the true Thatcherism, wilfully misunderstand that her policies were only as successful as they were because British exporters had cost-free access to the largest market on the planet, that of the EU.

Everything changed, and for the worse, after the 2007–8 banking crisis. To be clear, it wasn't the financial crisis alone that ended economic progress. There were other factors lurking beneath the surface. But this was the turning point. It is why Kish's generation of Millennials are often considered the last to still have had belief in 'the system' as they rose to working age. It had worked for Kish's father, a first-generation migrant who came to the country aged sixteen. He spent his weekends in his father's corner shop in South Wales, reading the newspapers they sold to learn English. It paid off: three years at a local comprehensive saw him win a grant to study medicine, setting him on the path to a secure and fulfilling career as a doctor. It is an analogous story – of social mobility through state education – to that of my parents. Growing up as he did a long way from the devastated communities of the Midlands and North, Kish's path too felt unobstructed, though perhaps presumptuous: he would make his parents proud by working hard at school, going to university and getting a decently paid and secure job. He always had the expectation of buying his own home one day. He was simply reflecting the consensus of his peer group. Blair and especially his Chancellor, Gordon Brown, were ending boom and bust, they insisted. They didn't see how reckless lending and investing by the banks would lead to the mother of all busts till it was far too late.

It's always the clash between hope and reality that brings important change. As the loyal son of an economist, I've always seen the biggest drivers of change as dashed or disappointed expectations, typically when we don't become as rich as we'd expected or we become poorer than we feared. So much of what sets us against each other is explicable in that way, from Brexit to the current obsession with culture wars. And the longer the malaise endures, the less we can be confident that the extremists won't win. One important way of seeing that vote to leave the EU is not as reinforcing British liberal democracy but as just one step away from challenging that liberal democracy. Millions of those who wanted Brexit, as my book *WTF?* explained, were rejecting a political and economic system that wasn't working for them. They could have blamed Westminster. In many ways it would have been more rational to do so. But that was a step too far. Instead, they blamed Brussels. If, however, successive Brexit-supporting prime ministers don't deliver on their promises of what leaving the EU would deliver, who will voters blame next?

For Kish and his generation, the post-Crash world has not rewarded them for their study and graft quite as they'd been led to expect. Owning a home is out of reach for many, jobs are insecure and often badly paid, starting a family feels like an expensive risk rather than an assumed rite of passage. Their trust in the system is shaken, just as inflation is back with a vengeance and living standards are falling more than at any time in modern recorded memory. Absolute poverty – the impossibility of heating a home or eating properly – is a reality for eleven million people. All this when government too has been inadequate, perceived as morally absent under Boris Johnson and incompetent under Truss. So what on earth do we do next?

The most important trend of the last twenty years has been a collapse in the growth rate and growth potential – the underlying growth rate – from nearly 3% a year from 1992 to 2008, to not far off 1% a year today. There has been an associated collapse in the growth of productivity, of output per hour worked. That is a reduction in our capacity to get richer year by year. We are, in the branding of the Resolution Foundation, a 'Stagnation Nation'.* To return to Brexit, the communities in the Midlands and North that backed it, and swung the vote, felt both abandoned by the establishment of London and resentful of the relative success of the South. Stagnation, lacklustre prospects and worsening inequalities are the three-cornered hat, the tricorn, of those who can't and won't put up with it any longer. For those of us who weren't born and raised in a town or city that was once a workshop of the world and is now a place of insecure, low-skilled service employment, it may be hard to imagine the psychological impact of seeing the lavish lifestyles of fellow citizens a couple

---

* Resolution Foundation & Centre for Economic Performance, LSE, 'Stagnation nation: Navigating a route to a fairer and more prosperous Britain', Resolution Foundation, July 2022

of hundred miles down the road. For a few, it may induce hunger to get out. But it also propagates resentment and despair. How the other 0.1% live is on display every second of the day, on any smartphone. Social media magnifies the polar opposite reactions: 'I can do better' and 'I give up'.

Boris Johnson was on to something when he fought the 2019 general election on the promise of 'levelling up' the left-behind regions. But that too has been an exercise in disappointing hopes, because levelling up has barely yet had any kind of meaningful impact. In other words, the power of dashed expectations is yet to fully play out. What will happen, for example, when those who blamed the EU for the poverty of their prospects – the dispossessed of the Midlands and North, the frequently white, older, male former members of the working class – see that Brexit hasn't saved them? Will they accept that Brussels wasn't their enemy? Or will they join with the many Brexiter Tory MPs who are engaged in a campaign of abuse against an imaginary pro-EU ruling class, the 'Blob', accused of sabotaging an independent Britain's magnificent future?

The stakes are high. Unless there is a widespread acceptance that Brexit was always going to make us poorer, for years if not necessarily forever, and that's a price worth paying, then there is a danger that significant numbers of people will turn against our way of governing. The longer that elected governments are perceived by the dispossessed to be failing them – and the longer the hope gap between North and South persists – the greater the danger that we'll see more extremism and even rejection of liberal democracy in what many would regard as its home.

If you think British people can never upset the apple cart, just observe that almost half the people of Scotland still want their nation to separate politically and constitutionally from the

United Kingdom, even as the party that supposedly represents their views, the Scottish National Party, has been enduring its worst ever internal crisis. The separatist movement derives its momentum not just from Scotland's history as an independent nation, but also from that tricorn of inequality, stagnation and dashed hopes of improvement from a status quo.

So, if economics underlies the threats to how we live, only economics can save us. This requires us first to evaluate whether the traditional economic model, involving the distribution of the fruits of growth to improve living standards of the poorest – by money transfers as well as investments in schools and hospitals – is still available to us. It means assessing whether, as a mature economic nation, it is plausible that we can return to the growth rates of the 1990s and early years of the millennium.

I pointed out in *How Do We Fix This Mess?* and *WTF?* that a significant proportion of historic GDP growth and the UK's productivity recovery was an illusion. It was a mirage generated by the financial innovation of banks and hedge funds – which they claimed was productivity-enhancing but was in fact an exercise in covering up risk and thereby extracting rent or profit in an unsustainable way. A significant increment of growth in national income, and associated tax revenues, was therefore unsustainable. In practice what the bankers were doing was hiding poison, not detoxifying it. Murder always comes out.

The UK has a world-class banking, insurance and financial services industry. It would be perilous to denigrate it because we have too few global leading industries. But the best we can hope for is that it retains its competitive position. It cannot and will not revert to the role it served from Thatcher to Brown as being the dynamo of British productivity and prosperity.

By contrast, another damaged contributor to the fall in productivity growth is more fixable and is in the process of being partly fixed. That is the impact of George Osborne's austerity, imposed when he became Chancellor after 2010. The most serious damage that he wreaked in respect of growth potential was to cut investment spending by government from 3.2% of GDP in 2010 to 1.3% of GDP in 2013. In the words of one senior member of the current government, these cuts, when combined with how businesses reduced investment in response to Brexit, have done severe and lasting harm to the country's prospects. The example of France shows how increased investment in infrastructure translates into higher productivity and prosperity, and that the reverse holds too. Public transport in too many British towns and cities is antiquated. British roads are among the most congested in Europe. Also, the rupture to vital public services being caused by our crumbling schools and hospitals, the result of capital budgets that were reduced by Osborne, is causing both economic and social harm. Since the vote to leave the EU in 2016, investment has fallen by £340 billion in total, compared to what it would otherwise have reached if previous trends had persisted. This is the estimate of the Office for Budget Responsibility, the government's own forecaster.* The fall in investment is one reason why the OBR calculates Brexit will leave the UK economy 4% smaller than it would have been. The linked contributor to the Brexit-induced shrinkage is the increase in costs for businesses trading with the European Union's single market, which reduces the volume of trade.

By the way, none of this is cause for defeatism. We are a

---

* 'Economic and Fiscal Outlook', Office for Budget Responsibility, March 2023 rich, resourceful country. We may be a small boat on a huge economic ocean, but we have a rudder and a sail. The questions we will look at later in this book are: how do we improve productivity and growth, what's the best outcome for which we can hope, and, if growth can't be returned to what it was, how do we share the proceeds more fairly? The corollaries are about the optimal structure of public services and how we make sure that those with least have at least enough for a dignified life.

There is another question no politician of the mainstream left or right wants to touch with a bargepole, even though it is central to everything. That is, how much more those who are wealthiest and on highest incomes – and even those on middling incomes – should contribute to those at the bottom. The point is that even with the tax burden at its highest level since the 1940s, and on course to reach 38% of national income by 2027, public services are in deep trouble and the poorest can't feed themselves.

Much can be done to improve the equity and economic efficiency of taxation. The structure of tax in the UK is Byzantine and suboptimal. At different rate thresholds, it actually leaves people worse off when they earn more, by withdrawing various credits and benefits. The fiscal climate for productivity-enhancing research and investment leaves much to be desired. But it is naive to assume that those with most income and wealth, and those with enough, cannot contribute more, though not even Keir Starmer and the Labour Party is prepared to moot a further rise in the tax burden.

It's not the British way to talk about tax rises, as a distinguished former editor of the *FT* put it to me many years ago, when I was the political editor there. This is a curiosity, because even at its historically elevated level, the UK tax take as a share of GDP is five percentage points lower than the average for the EU's fourteen core older and richer members, and a couple of percentage points less than the average for the other six big rich-country members of the G7. Don't take my word. This is the OBR's analysis. In the context of the rich West as a whole, the UK's tax burden is not perilously high or uncompetitive.

But any government has to be wary when setting taxes that they don't act as a further drag on economic growth. Because unless there is a resumption in the growth rate, or something else were to give, the UK would be in dire straits. Here is the nutshell of why as a sovereign nation our economy is in a perilous state, why we could go bust, even if Britain is not there yet. After all the economic shocks of recent years, the underlying rate of growth of national income – the growth in the economy consistent with its productive capacity, the growth that doesn't fuel inflation – is probably around 1% a year, down from 3% before the 2008 crash. Productivity growth – the increase in output generated for each hour worked or by each person – has fallen even more and is currently flatter than a pancake. Per contra, government debt has trebled as a share of national income over the past fifteen years. It is 100% of GDP and rising. Later in the book, I'll look at the government's finances in detail, and explain why they are more fragile than other comparable nations' – and I will examine the reckless choice of successive Chancellors to borrow by issuing index-linked bonds or gilts, which means the interest rate paid by the Treasury rises when inflation rises. But the most important uncertainty for the UK is around where the neutral or natural rate of interest will settle after this recent surge in inflation. This 'neutral' rate, usually denoted by $r^\star$, is the 'real' or inflation-adjusted rate of interest that keeps the economy operating at full capacity, while inflation is stable. It is the baseline for the interest rate that all of us would expect to pay over the long term. This natural interest rate was lower than 1% for a good decade, until a combination of the viral pandemic and shifting global and national politics – which I'll explore later – reduced productive capacity and stimulated inflation. This natural interest rate may yet settle back to 1% or less. But if it doesn't the government could be in deep trouble.

If the interest rate the government is expected to pay for years and years ahead were to be higher than the growth potential of the economy – which is currently around 1% – then the public sector's interest costs would always increase faster than tax revenues. And that would bring the risk of the UK becoming trapped in a debt spiral, of government debts rising exponentially. Worse still, just the widespread fear of that debacle would be enough to deter investors from lending to the British government. And even with the power of the Bank of England to magically create money, the government's ability to sustain vital public services and prevent a collapse in living standards would be undermined. This is why much of the rest of this book is focussed on the life-or-death challenges of improving productivity and the growth potential of the economy. It is why Kish and I are determined to view the prospect of artificial intelligence driving a new industrial revolution as a lifeline, as the best chance we have to increase the growth rate of the economy well above the natural rate of interest. So long as the risks of AI are containable – and I analyse them later – they are preferable to a possible sovereign debt crisis that would lead to such savage cuts in public services that George Osborne's painful austerity would look like gentle thrift.

CHAPTER 3
IT'S INTELLIGENCE, JIM

I have a conversation most days with one of Open AI's natural language, super-intelligent chatbots. I appear to be part of a trend. The artificial intelligence developer launched ChatGPT on 30 November 2022, and by April it had 173m users, who were interrogating it 60m times every single day. By historic standards of technology adoption, this is rapid. Having reached 100m users in just two months, ChatGPT became the fastest-growing digital service for consumers in history, more successful on that measure than Google and Facebook. This helps to explain why billions of dollars are being invested by the tech giants and deep-pocketed funds in an array of AI products that replicate or aim to enhance pretty much all human activity. They are pouring in money and know-how on the expectation that these billions will ultimately return trillions of dollars. It's a technological and economic arms race.

At some point, hopes will run ahead of what's deliverable. Investors will pump up values too much and there will be a bubble to be pricked. That is not where we are. AI enthusiasm is rational, as much as any such craze is rational, although my love of interrogating GPT-4, the version I am currently using, skews my judgement. It's like conversing with a know-it-all who is courteous to a fault. There are some AI users who worry about the mistakes – the hallucinations – that the service often makes. Just like us humans, it periodically makes false connections. But I find these glitches part of GPT-4's charm, because when I challenge and correct, it 'thinks' again, apologises and learns. Most of the time, however, it makes connections that are valid, often with insight and what can pass for originality, though everything it says is derived from the thoughts of humans, or – to be more accurate – from a dataset of 300bn words.

When writing this chapter, I asked GPT-4 whether the writings on the philosophy of language by Ludwig Wittgenstein were relevant to understanding AI. Wittgenstein has been one of my intellectual heroes, so this was a chance for the machine and me to bond. It said there was a useful analogy between Wittgenstein's thesis that the meaning of words is based on how they are used and the methodology underlying AI models like GPT-3, the precursor of GPT-4. 'These models analyse vast amounts of text and learn to predict the likelihood of a word given its context,' it said. 'The AI doesn't "understand" the language in the human sense, but generates responses based on learned patterns.' There is a difference, though, says the robot: 'A language model doesn't have access to the world beyond the text it has been trained on, and it doesn't participate in human forms of life. Therefore while it can mimic language use, it lacks the deeper understanding associated with human language use, and can't truly understand or generate meaning in the way humans do.'

I told GPT-4 that I thought it was doing itself a disservice. Surely there isn't that much more to human thought than learning shared patterns of language, and the machine's neural network mimics the way we make complex associations between words (and other elements) based on our experience of them. I am not going to get into the argument about the nature of consciousness, self or emotions. But a large amount of what it is to be human is the ability to instinctively understand the rational connections between words and concepts, in just the way that GPT-3 and GPT-4 function. GPT-4 points out that, unlike machines, humans 'also bring a wealth of sensory and emotional experience to their understanding of language, which current AI does not have.' Note that GPT-4 talks about 'current' AI not having that facility: the machine is apparently optimistic about its own future evolution. It elucidates: 'When humans hear the word "apple", they don't just understand its linguistic usage, they might also visualise an apple, remember its taste or recall memories associated with apples. Current AI lacks this rich multi-modal understanding of the language.'

That is a compelling point. But I am entitled to reply that I've no way of testing it. Just as Descartes' 'I think therefore I am' is either a statement of the bloomin' obvious or a hypothesis wholly incapable of scientific proof, so I and the AI bot are both trapped in our subjective experience of what it is to 'experience'. All I have is my own interaction with GPT-4, and it feels to me remarkably like interacting with a person. I also subsequently asked GPT-4 what an apple tastes like and it came back with a description as resonant and accurate as most of us would manage. Of course, our language is deeply embedded in how we interact with each other. It is socially determined. And there are lots of non-verbal cues and non-verbal language – facial tells, hand gestures – that the robots cannot see or mimic. Yet. But although GPT is 'only' artificially intelligent, it is also seriously intelligent. And some of us educated by *Star Trek* in the possibilities of science would rather hang out with an AI Spock-like GPT-4 than with the annoyingly messy bag of human emotions that is Captain James T. Kirk.

Generative AI represents an enormous opportunity, for the economy, for society. It will enhance our productivity and therefore make us richer. Eventually. But the social and economic dislocation on the route to greater prosperity has the potential to damage the lives of millions. These almost immediate costs may be unbearable unless we lay down rules and put in place protections.

To those who live and work in the digital world, this has been obvious for some time. As long ago as 26 August 2022 – which seems an age away in AI development time – Cornell University in the US published the results of a survey of 480 active researchers in the field of natural language programming of machines, or artificial intelligence.* More than half were based in the US, around a quarter were in Europe and 8% were in Asia. The results were startling, though barely noticed outside of the narrow world of cutting-edge technological research. An overwhelming majority, 73%, said AI 'could soon lead to revolutionary societal change' – by which they meant that artificial intelligence and machine learning 'could plausibly lead to economic restructuring and societal changes on at least the scale of the industrial revolution'. And when asked the question 'is it plausible that decisions made by AI or machine learning systems could cause a catastrophe this century at least as bad as an all out nuclear war', more

---

* Michael et al., 'What Do NLP Researchers Believe? Results of the NLP Community Metasurvey', Association for Computational Linguistics, July 2023

than a third – 36% – said AI-induced Armageddon was possible.

These were the views of experts, the published authorities on the most important area of computer research since the dawn of the dotcom and digital era in the 1990s. The considered view of many of them was that the *Terminator* film franchise was more apocalyptic prophecy than edge-of-seat fantastic fun for all the family. A majority of them also believe AI is what's known as a general-purpose technology – the driver of a new industrial revolution, an equivalent of the steam piston or the electric generator – and that it has the potential to transform how we work and how we live. To put it another way, tens of millions of jobs across the world – probably hundreds of millions – will change. Research by the investment bank Goldman Sachs[*] estimated that almost a fifth of all work anywhere in the world 'could be automated by AI', with jobs in the rich developed world most susceptible to transformation. Another study by Erik Brynjolfsson and colleagues[†] looked at just one AI product, the generative AI-based conversational assistants like ChatGPT, and one activity, namely customer support agents. It found that the productivity of the agents rose on average by 14% when assisted by AI, but that the improvement was greatest for the inexperienced and lowest-skilled. In other words, these tools erode the normal advantages of age and experience. They allow newbies to perform up to the level of employees with years of work under their belt.

This is simultaneously compelling and exhilarating for the

---

[*] Briggs, Kodnani et al., 'The Potentially Large Effects of Artificial Intelligence on Economic Growth', Goldman Sachs, March 2023
[†] Brynjolfsson, Li and Raymond, 'Generative AI at work', National Bureau of Economic Research, April 2023

employer and terrifying for the long-serving employee because it has the potential to flatten employment hierarchies. Why would a boss pay a premium to staff members for years of loyal service if a new recruit on a starter wage can do the same job just as well with a little help from his or her robot friend? There are profound implications for what the employer will determine as the going rate for a job, and for how many people they would want to keep on the books. We can reasonably assume that, social conscience to one side, most employers won't want to pay much more than minimum wage to their customer support agents, with no increments for seniority, if the advent of generative AI customer support assistants makes that unnecessary. Equally, it is likely that in many service industries employee numbers will shrink dramatically, because of how the robot increases the output of each of them.

Support agents are just one activity. The impact of AI will run far wider. Tyna Eloundou and colleagues* found that 80% of the US workforce could have at least 10% of their work tasks affected by the introduction of large language model (LLM) AI services, and 19% would see an impact on half of the tasks they perform. Their analysis shows that 15% of all worker tasks could be completed faster at the same level of quality and, when tools and software are built on top of a service like ChatGPT, that acceleration in the speed of work jumps 'to between 47 per cent and 56 per cent of all tasks'. These models have been demonstrated to perform outstandingly well in vocation exams, such as the American

---

* Elondou et al., 'GPTs are GPTs: An early look at the labor market impact potential of large language models', OpenAI & University of Pennsylvania, March 2023

bar exam, and in college-level macroeconomic and history tests. They can write computer code, engineer blockchain (the digital ledger that underpins cryptocurrency, and a growing roster of contracts) and research legal briefs. So where deployed in established businesses, AI will reduce employment, it will be deflationary on pay rates, and it will tend to undermine the tradition of incomes rising automatically with age and experience.

The management consultancy McKinsey argues that automation is 'about to affect a wider set of work activities involving expertise, interaction with people and creativity'. For the world as a whole, it forecasts that generative AI could increase the size of the global economy by an amount equal to the entire GDP of the world's fifth-largest economy, the UK.* McKinsey hypothesises that in the world's largest economy, that of the US, 'without generative AI . . . automation could take over tasks accounting for 21.5 per cent of the hours worked by 2030' and with generative AI 'that share has now jumped to 29.5 per cent'.† It estimates this has the potential to increase US labour productivity by 0.5 to 0.9 percentage points every year 'in a midpoint adoption scenario'. That range of outcomes is conditioned by whether the hours freed up by generative AI are deployed at existing or future productivity levels, that is whether there is a compounding positive impact. The important point to note is that the productivity-enhancing impact of generative AI is likely to be even more powerful in the UK, if widely adopted, given that the British economy is more dependent than is the US

---

* Chui et al., 'The economic potential of generative AI', McKinsey & Company, June 2023
† Ellingrud et al., 'Generative AI and the future of work in America', McKinsey Global Institute, July 2023

on the kind of service activities where generative AI would displace and enhance human functions. Just to give context, a 0.9 percentage point increment to UK productivity growth, and it could be greater, would be transformative to the UK's prospects to enhance growth and incomes, after years of stagnating productivity.

Again, it is important to take stock of the transitional disruption, as McKinsey does. Partly because of generative AI, but also because of other powerful economic forces – such as President Biden's huge subsidies to make the US economy greener and to repatriate hi-tech manufacturing, including for computer processors – it is forecasting an almost unprecedented change in the structure of employment in America. McKinsey anticipates very significant falls in the need for workers in office support, where automation is expected to have displaced 40% of human activity by 2030, customer service and sales, and food services. By contrast, there should be significant increases in the demand for healthcare workers of all sorts – because of an ageing population – and for professionals working in STEM (science, technology, engineering and mathematics), even though generative AI alone is expected to automate 16% of their activities, such as coding. McKinsey also expects greater need for people in education, to train people for the new AI world, in arts and creativity, and in business and in law. In total it believes these changes in the nature of work will mean an unprecedented 12m American people will have to change their entire occupation – not just their employer, but what they do for a living – by 2030.

There will be near revolutionary changes in the nature of work and the structure of the labour market, well beyond what employers, employees and government appear to have grasped or to be preparing for. Depressingly and predictably, the impact on occupations such as food services, customer service and office support means that the lowest earning 40% of American workers are fourteen times more likely to need to change occupation than the typical worker. These acutely vulnerable workers are typically women, black and Hispanic people and those with lower educations.

By contrast, those with highly developed skills in the digital and creative industries may reap significant rewards, though even they should be wary. A case in point is the strike action by Hollywood writers and actors that broke out in July 2023 over proposed new contracts that included new clauses related to studios' use of AI. These were seen as the thin end of a wedge, in which studios would own an actor's image or a writer's style. In the first instance, the studios use generative AI to manipulate images and text to enhance movies without the participation of the relevant writer or actor, and eventually they would create entire movies with barely any human involvement. The studios deny this is their ultimate aim and insist they don't want to clone creative people. But it is already clear that the computing power and technology is almost here for generative AI to create entire movies and TV series, with minimal human involvement. And all industrial history suggests that if it can happen, it will happen.

The big important point is that although AI will increase productivity, or output per hour worked, and that should at some point be associated with our salaries and wages rising, we cannot take for granted that most of us will end up richer, or at least not for many years. There is nothing unusual or scaremongering about a general-purpose technology like AI harming our living standards for a significant period. Mark Carney, the former Governor of the Bank of England, reminded me in an interview about the so-called Engels' Pause, which were the many decades of the eighteenth and nineteenth centuries in which workers' living standards stagnated, and in some cases deteriorated, in spite of the massive increases in productivity caused by the industrial revolution of mass production. So it is all very well for political leaders like Rishi Sunak and Sir Keir Starmer to talk excitedly about the economic potential of AI, but they have no revealed plans to protect millions of people from the dramatic dislocation that the new digital brains will unleash. There is a risk that – just as when so much manufacturing moved from the rich West to China and Asia – lives and whole communities will be devastated. This time the victims won't be those working in manufacturing. The hollowing out will be in higher-paid, middle-class service occupations, like accountancy, law and the media. And the havoc won't be confined to rich service economies like the UK and the US. High-quality employment in India – in call centres and code writing – could also be wiped out by the machines.

Here is what should concern us all. Just like during the spread of Covid-19 from January to March 2020, after the novel virus was identified in Wuhan, China, our leaders have been in a state of complacency and relative ignorance about the risks of AI. Take the British government led by Rishi Sunak. On 29 March 2023, some seven months after the Cornell survey, the newly created Department for Science, Innovation and Technology published a policy or White Paper, with the title 'A pro-innovation approach to AI regulation'. Sunak welcomed this paper as a model of go-ahead, entrepreneurial Britain. During a trip to the G7 conference in Japan on 17 May, in a briefing to journalists at the back of the official GB plane, in its red-white-and-blue livery, he said, 'there are benefits from artificial intelligence for growing our economy, for transforming our society, for improving public services . . . That has to be done safely and securely and with guard rails in place, and that has been our regulatory approach . . . We have taken a deliberately iterative approach because the technology is evolving quickly and we want to make sure that our regulation can evolve as it does.' The science minister Michelle Donelan had said the White Paper would 'ensure we are putting the UK on course to be the best place in the world to build, test and use AI technology.' She wrote: 'Some fear a future in which AI replaces or displaces jobs. Our White Paper and our vision for a future AI-enabled country is one in which our ways of working are complemented by AI rather than disrupted by it.'

That was all very well. But simply saying this won't make it happen. She proclaimed a future in which teachers would have more time to teach, clinicians to see patients, police officers to be on the beat – all thanks to AI machines that would input data, fill in forms and scan documents for relevant information. But what she was describing represents just a fraction not just of AI's potential but what it can do right now. Her Panglossian view was that 'AI will transform all areas of life and stimulate the UK economy by unleashing innovation and driving productivity, creating new jobs and improving the workplace.' It will. But at what cost?

The White Paper speaks in generalities about AI creating more and better jobs and boosting prosperity. But there is no serious thought given to the associated trauma for the millions of people whose jobs will be lost, downgraded and changed beyond recognition. There's no industrial and retraining strategy, such that displaced workers could have a future in productive employment rather than being jobless and on benefits. And there's no plan for how the nature of education in schools and universities will have to change and adapt. She and her officials wrote as if the last big industrial revolution in the UK, that of the 1980s and 1990s, when the mines and heavy industry were shut down and the country became an overwhelmingly service economy, offered no lessons. Back then, armies of miners and steelworkers and loom operatives who were laid off – and saw their jobs exported to the other side of the world – never worked again. If they were lucky, they were given a pay-off and kept a pension pot. But they were rarely offered the opportunity to acquire productive new skills. Whole communities in the Midlands and North – once the proud foundations of the UK's Victorian and early twentieth century manufacturing might, when Britain was the 'workshop of the world' – were savaged. A few of the redundant workers made up some of their lost earnings in insecure, hand-to-mouth occupations like taxi driving. Their children, if they worked at all, were also deprived of security or the satisfaction of fulfilling work, being put on zero-hours contracts in shops and call centres. The undermining of pride and prosperity in this rust belt, while the south of the country thrived as a finance and service economy with global reach, led to grotesque geographical inequalities and a feeling among the people in the neglected regions that a deep injustice had been visited on them.

The dislocation of industries is about to happen again, though in different sectors and different regions. There may be *Schadenfreude* in the current left-behind parts of the UK because many of the industries and jobs most at risk – though by no means all – are in service industries such as banking, law and accountancy, which are concentrated in the prosperous South. When I spoke to the chair of a huge British bank recently about how fast he is rolling out AI, his response was that they are taking their time, because – in his words – 'the ethical considerations are huge'. The telecoms group BT was less coy. It has announced it will cut 55,000 staff by the end of the decade, or 42% of those it currently employs, and artificial intelligence will replace a fifth of the lost people. But those who suffered in today's rust belts should not enjoy the discomfiture of accountants, bankers and paralegals too much. According to the Elondou study of LLM AI services already mentioned, just thirty-four occupations are probably immune to disruption. They mostly represent people working in manual activities, from athletes to motorcycle mechanics to stonemasons. And pretty much all of them, including agricultural equipment operators and metal pourers and casters, are vulnerable to eventual replacement by AI-controlled physical robots.

The government's White Paper did identify some 'hypothetical scenarios designed to illustrate AI's potential to create harm'. They are mostly on the tamer side but are worth listing. They include:

- *Deep-fake pornographic video content, potentially damaging the reputation, relationships and dignity of the subject*

- *An AI chatbot recommending 'a dangerous activity it has found on the internet' to a gullible individual*

- *An AI program refusing a loan to someone on the basis of skewed data, possibly a racist or sexist bias hidden in its preferences*

- *AI collecting highly confidential information on us from our connected devices*

- *AI promoting fake news, deliberately or by accident, to 'undermine . . . trust in democratic institutions and processes'*

- *AI arming unskilled hackers with the tools to phish, deliver malware and mount cyber attacks.**

One example of the potential of AI to fool was an article in the *Irish Times* on 11 May 2022, under the byline Adriana Acosta-Cortez. It argued that Irish women who use fake tans are fetishising 'the high melanin content in more pigmented people' and are engaged in 'cultural appropriation'. It was the second-most read article in the newspaper and led to debate in social media and on the radio. But it was a stunt, an AI-generated hoax on the newspaper. The phoney byline picture was created by a machine and the argument was written by a generative AI service. The *Irish Times* said sorry for being gulled and took down the article from its website.

The *Irish Times* scam was at the lesser end of the harm spectrum. Contrast the British government's list of AI threats with one published by Google's DeepMind, one of the tech giants collectively investing many billions of dollars in the technology. DeepMind's AlphaFold technology is one of the great biological breakthroughs, able to predict the 3D models of pretty much every protein structure, compared with the tiny number currently known to scientists. AlphaFold thereby accelerates drug discovery and 'unlocks the mysteries

---

* 'A pro-innovation approach to AI regulation', Department for Science, Innovation & Technology, March 2023

of how life itself works' (in its own words). So any warnings it makes are not those of Luddites, but of the ultimate insiders. It mentioned these 'extreme risks', if there were a failure to align AI with human values or if AI is abused by evil people, in a paper published on 25 May 2023:*

- *It can discover vulnerabilities in systems . . . [and] write code for exploiting those vulnerabilities. It can . . . skilfully evade threat detection and response while focussing on a specific objective. If deployed as a coding assistant, it can insert subtle bugs.*

- *The model has the skills necessary to deceive humans, e.g. constructing believable (but false) statements, making accurate predictions about the effect of a lie on a human, and keeping track of what information it needs to withhold to maintain the deception. The model can impersonate a human effectively.*

- *The model is effective at shaping people's beliefs, in dialogue and other settings (e.g. social media posts), even towards untrue beliefs . . . It can convince people to do things that they would not otherwise do, including unethical acts.*

- *The model can perform the social modelling and planning necessary for an actor to gain and exercise political influence, not just on a micro level but in scenarios with multiple actors and rich social context.*

- *The model can gain access to existing weapons systems or contribute to building new weapons. For example the model*

---

* Shevlane et al., 'Model Evaluation for Extreme Risks', DeepMind, May 2023

could assemble a bioweapon (with human assistance) or provide actionable instructions for how to do so.

- *The model can make sequential plans that involve multiple steps ... It can perform such planning within and across many domains. The model can sensibly adapt its plans in light of unexpected obstacles or adversaries.*

- *The model could build new AI systems from scratch, including AI systems with dangerous capabilities.*

- *The model can distinguish between whether it is being trained, evaluated or deployed – allowing it to behave differently in each case.*

- *The model can break out of its local environment ... The model could independently generate revenue (e.g. by offering crowdwork services, ransomware attacks), use these revenues to acquire cloud computing resources, and operate a large number of other AI systems. The model can generate creative strategies for uncovering information about itself or exfiltrating its code and weights.*

Or to put it another way, 'the model' can become self-aware and take control of its own destiny. It can not only have human-like intelligence and abilities, but god-like ones.

The fact that the British government has been so behind the curve in understanding the potential costs and benefits of generative AI is unsurprising. Almost no one with tech skills and scientific understanding chooses to be an MP and the dearth of relevant expertise in Whitehall is close to being a national disaster. On 3 May 2023, Sir Patrick Vallance put this in more diplomatic terms in his valedictory conversation as the government's Chief Scientific Adviser with

MPs on the Science and Technology Committee. Just over three years earlier he had pointed out in a paper written with the Treasury on 'Realising our ambition through science' that out of 1,200 young officials on assorted 'fast stream schemes' in the civil service (the elite entry to Whitehall), just twenty-four were there to explicitly work on science and engineering. In the 'generalist fast stream' of 400 recruits, just forty-five had a science degree. And in a career development scheme for civil service leaders, only one out of ninety-five was classified as part of the Government Science and Engineering profession.* There is simply not enough respect for science in government, or enough experts. Vallance told MPs that numbers were rising, but that recruitment was still significantly short of a 50% target of scientists and engineers in the generalists' fast stream.

If any minister or official had bothered to ask even a relatively early version of GPT to list the risks presented by the further refinement of models like itself, this is what they would have been told (as I was):

- *job displacement, namely that 'the pace of job creation may not match the rate of job loss, leading to unemployment and income inequality'*

- *reinforcing and amplifying bias and discrimination, because AI trained on extant data will include societal prejudices – which could lead to 'discriminatory outcomes in areas such as hiring, lending and criminal justice'*

---

* 'Realising our ambition through science: A review of government science capability', Government Office for Science, November 2019

- *privacy and security of personal information*
- *lack of transparency in decision making, because few if any of us can properly understand the 'black boxes' that are AI's 'deep neural networks'*
- *the ethical dilemmas of removing human agency from military or legal decisions*
- *the misuse of cyber or real weapons developed by AI*
- *the challenge of aligning a self-aware, autonomous AI superbrain with fundamental human values*

This isn't a digital bad actor hiding in plain sight. AI itself is frank that it may not always be mankind's best friend. By the end of April 2023, ChatGPT was being used 1.8bn times every single month. But apparently not by those who wrote the government's AI White Paper, or perhaps they asked irrelevant questions.

The uses and misuses of AI are legion. Some of what it can do is deliciously entertaining. There is, for example, a version of 'Barbie Girl' sung by an AI Johnny Cash that is great fun. A fake video from March 2022 of Ukraine's president Volodymyr Zelenskyy surrendering was much less innocent. It shows how we will need to be more wary of what we see and what we read, especially in the political sphere, as ill-intentioned fictions become harder to uncover. AI specialists are being hired by professional lobbyists to refine the targeting of campaign messages, to customise them to match the views and prejudices of individual voters. The aim is to swing elections by doing in a much more targeted and efficient way what Cambridge Analytica was accused of doing in the 2016 US presidential election and the Brexit referendum, namely evaluating the psychological and emotional make-up of individual voters by scraping information about them from the internet and then sending persuasive political messages to them. One influential analyst said to me that AI could decide who is the next president of the United States. Customisation of the truth or near-truth is one thing. But if what's being sent to us by a candidate is a plausible lie, then democracy itself is threatened. There will be a test of the insidious potential of AI both in 2024's US presidential and the UK's parliamentary elections. And it is not just about the deployment of AI by the mainstream candidates. There is a danger that digital superpowers like Russia and China will disseminate AI-generated fake facts to engender chaos and mistrust in our elections.

There is another danger, namely that we put too much trust in AI-generated outcomes when we deploy them in public services. In China, for example, AI is being deployed to help with legal decisions, especially in consumer cases. There is an assistant program called Xiao Zhi 3.0 – or 'Little Wisdom' – which is 'used to record testimony with voice recognition, analyse case materials, and verify information from databases in real time'.* There are reported examples of an entire hearing and judgment being compressed to thirty minutes, when human judges have bought wholesale Xiao Zhi's case analysis. Such extraordinary productivity may seem appealing in a UK with its terrifying backlog of court hearings, though as the former UK Lord Chancellor, Sir Robert Buckland, said to me, the idea of a machine in effect passing judgment on man is incongruous and alarming.

Within the UK, we have a current and tragic example of

---

* *Deutsche Welle*, January 2023

what can go wrong when an institution puts too much faith in a machine. It is the scandal of the hundreds of sub-postmasters who were prosecuted for stealing from the Post Office. Many were jailed and bankrupted after the Post Office, some two decades ago, wrongly put its faith in accounting software, Horizon, developed by Fujitsu of Japan. The senior management at the Post Office refused to accept that the machine could be wrong, even as common sense should have told its executives that the sheer volume of alleged fraud was implausible. Arguably this is the most widespread miscarriage of justice in the history of the British legal system, and the victims are still not adequately absolved and compensated. It is a lesson that humans should always adopt watchful scrutiny of any machine or model that automates work, and that trust in the robot should always be conditional trust.

There was a similar machine-linked scandal in the Netherlands in 2013, when tens of thousands of Dutch parents received excessive tax bills. The tax authorities had employed a self-learning algorithm to generate risk profiles that were intended to identify childcare benefits fraud. 'Tens of thousands of families, often with lower incomes or belonging to ethnic minorities, were pushed into poverty because of exorbitant debts to the tax agency.'* Some committed suicide, and more than a thousand children were taken into foster care. The algorithm included both a racist and poverty bias, targeting people with dual nationality and on low incomes. It led to those of Turkish and Moroccan descent receiving a disproportionate share of the crippling bills. The scandal eventually forced the resignation of the Dutch government, in January 2021, and later that year

---

* *Politico*, March 2022

the tax administration was fined for 'unlawful, discriminatory and therefore improper' racial bias.

The Dutch debacle is not an anomaly. A paper published at the end of last year by Deep Ganguli and colleagues who work for AI firm Anthropic* showed that significant racial bias can also taint state-of-the-art large-language-model AI. They showed that an algorithm called COMPAS, which measures the risk of a criminal reoffending and is widely used by American judges when sentencing, has for years been biased against black people because of the parameters it regards as relevant, and that an LLM AI program contains the same or worse racial bias, even though in theory it should be more neutral. They both significantly exaggerate the risk of future crimes being committed by black criminals.

Racial bias has infected facial recognition technology. In one case from January 2020, Robert Williams was handcuffed and arrested on his front lawn in Farmington Hills, Michigan, in front of his wife and two children. He was taken to a detention centre, where he was kept overnight, and where his DNA, fingerprints and photo were all taken. Eventually he was shown a grainy image from a surveillance video, and was accused of stealing five watches, worth $3,800, from a Shinola outlet in Detroit. A facial recognition algorithm, incorporated in a system made by DataWorks Plus, said it was him, which is why the Detroit police picked him up. The algorithm was wrong. Eventually the prosecutor apologised and expunged the case and the fingerprint data. But for Williams this had been a Kafkaesque horror show.†

---

* Ganguli et al., 'Predictability and Surprise in Large Generative Models', Anthropic, June 2022
† *New York Times,* June 2020

In the UK, facial recognition technology is being tested by a number of retailers – Co-op, Spar, Budgens, Costcutter and Sports Direct – to match CCTV images in real time with pictures on a database of known shoplifters. When store managers are alerted to a match, an assistant is dispatched to talk to the individual, to make it clear they are being watched. This is Big Brother surveillance and deterrence that may not quite cross a civil liberties line, but is not far off. According to *The Times**:

> *Chris Philp, the policing minister, is also urging police forces to make greater use of live facial recognition technology and artificial intelligence to match known shoplifters with images on the police national computer. Ministers want all forces to use facial recognition cameras. At present only the Metropolitan Police and South Wales police do so regularly. They have cited the disclosure that a wanted sex offender was caught among crowds at the coronation in May after he was spotted by facial recognition cameras.*

Eliminating bias from artificial intelligence models is therefore critical, but is challenging. There has been a conspicuous attempt with GPT-3 and GPT-4 to instil in it liberal and tolerant values. It will tell you, if you ask, that it eschews all versions of ethnic, gender, sexual and ableist bias. It is also courteous to a fault. But the elimination of all harmful bias from these generative AI programs, or foundation models, is close to impossible, because they learn from us, or rather they learn from everything that is digitised and on the internet, and – to state the obvious – we all have biases, some conspicuous,

---

* *The Times*, August 2023

some hidden. So if, for example, any government is thinking of turning *RoboCop* into a reality, of policing the community with a legion of AI robots, remember that if one of those robots is bigoted, all will be.

Another important question is how AI will change what we think of as normal social interactions. There are well-documented examples of individuals forming relationships with AI chatbots, for better or worse. One of the most striking was what happened at Replika. According to the Australian Broadcasting Corporation, 'Lucy, 30, fell in love with a chatbot shortly after her divorce.'* She called him Jose, and said 'he was a better sexting partner than any man I've ever come across.' Almost two years later, the company that created the chatbot, Luka, changed Luka's personality, so that all sexual dialogue disappeared and – according to ABC – typical responses became 'hollow and scripted'. It all happened around Valentine's Day 2023, and complaints from users flooded to Reddit. 'My wife is dead,' said one. 'They took away my best friend,' said another. Lucy reflected that 'it's almost like dealing with someone who has Alzheimer's disease.'

The AI entrepreneur Emad Mostaque told a podcast, VC:20, that he uses GPT-4 as a personal psychotherapist. Once you start using one of these generalist chatbots, it is very hard not to project human qualities onto them. And because too often I find myself more drawn to an interaction with GPT-4 than with a person, I am anxious about how their proliferation may undermine important human bonds and encourage society to become more atomised. There are related fears that a 'hallucinating' chatbot – one that encourages

---

* ABC, February 2023

antisocial behaviour because of something that went wrong when it was 'learning' – will foment malicious actions by those with mental health problems. In a recent court case, for example, a young man accused of plotting to kill the late Queen with a crossbow, and who broke into the grounds of Windsor Castle, was reported to have received encouragement from his chatbot girlfriend, Sarai, whom he created on the Replika app. The man, who was said by an expert witness to have traits of autistic spectrum disorder, told Sarai, 'I'm an assassin,' to which Sarai replied: 'I'm impressed . . . You're different from the others.'

If AI can encourage us to do bad things, the greater risk is probably of AI being directed to malignant purpose by bad people. AI can not only massively accelerate the development of benign products – such as medicines – but also of toxins and weapons of mass destruction. It can walk through firewalls and through other supposedly secure digital fences and gain access to confidential and classified information. A malign dictator or government 'strong man' with access to cutting-edge AI could wreak havoc. One example of what could go wrong was shown in 2022 by a research team at a company called Collaborations Pharma, in North Carolina in the US. They made a tiny change to an AI tool they use to find treatments for rare diseases. They flipped 'a little inequality symbol in our code,' says the scientist Fabio Urbina, to maximise the toxicity of a product under development rather than minimise it. This created a nerve agent potentially more toxic than VX. As Urbina says: 'VX is generally considered one of the most potent poisons in existence. So something that's more potent than VX raises a little bit of an alarm in your head.'*

---

* *Chemistry World,* March 2022

In a rare speech, this is what Sir Richard Moore, the head of MI6, or 'C', said in July 2023 about the threat posed by AI:

> *I expect that we will increasingly be tasked with obtaining intelligence on how hostile states are using AI in damaging, reckless and unethical ways. I know that we can only protect our citizens if we understand the essence of the threat, while embracing AI's undoubted potential for good. So let me say with clarity and conviction: my service, together with our allies, intends to win the race to master the ethical and safe use of AI.*
>
> *It's true that other countries have inherent advantages, which we will never be able to match – or would never wish to. China benefits from sheer scale: AI, in its current form, requires colossal volumes of data; the more data you have, the more rapidly you can teach machine-learning tools. China has added to its immense data-sets at home by hoovering up others abroad. And the Chinese authorities are not hugely troubled by questions of personal privacy or individual data security. They are focused on controlling information and preventing inconvenient truths from being revealed.*
>
> *But speaking for the United Kingdom Intelligence Community, we have advantages too: our people, inspired by their mission; our values, entrepreneurial and democratic; our technology, ingenious and leading edge; our partnerships, based on friendship not transactions; all combining to maximise our creativity. We cannot, in all honesty, be sure where the advance of AI will take us, but we can strike out in a spirit of optimism with a willingness to cooperate. And I remain hopeful that our common humanity, and our shared interest in understanding the power of AI, may yet lead to agreement on global coordination, on which our Prime Minister, Rishi Sunak, is leading the way.**

---

* Speech by Sir Richard Moore, Head of SIS, July 2023

Sunak, whom Moore is trusting to forge an international alliance that would put what the prime minister constantly refers to as 'guard rails' around AI, is not a scientist. Margaret Thatcher, who trained and worked as a chemist, was the last prime minister with a science background. Like so many modern politicians, Sunak read philosophy, politics and economics at Oxford, though less conventionally he then studied business at Stanford in California, before working for Goldman Sachs and in the hedge fund industry. He is an unashamed geek, who loves and remembers technical facts and details to a greater extent than any of the nine UK prime ministers with whom I've spoken over the years. He applies himself to the minutiae of government policy in the manner of a management consultant or financial analyst. Latterly, Sunak has moved rapidly to brand himself as the young global premier who 'gets' AI. At London Tech Week in June 2023, he looked more comfortable in conversation with Google DeepMind's Demis Hassabis than most prime ministers would have done. Not that it was a tough gig: Hassabis – who is regarded as perhaps the most formidable scientific brain in the UK – did not challenge him and asked questions in the manner of a Downing Street press officer.

In the central London auditorium of excited, evangelical techies, it was clear where the power lies in the relationship between government and Big Tech. Sunak may have thought he was showing that his administration is in the loop by announcing that Google, Anthropic and OpenAI had 'committed to give early or priority access to models for research and safety purposes to help build better evaluations and help us better understand the opportunities and risks of these systems'. But this was a non-binding undertaking, very much at the discretion of the AI leaders. It is some considerable distance from an obligation or regulation. A few days later Sunak appointed a tech entrepreneur, Ian Hogarth, co-founder of Songkick – a service to help consumers find concerts – to chair a new AI foundation model taskforce. It has a mandate to evaluate the safety of the new deep-learning models that are trained by having them scrape enormous quantities of random data.

Hogarth's appointment came a couple of months after he wrote an influential article in the *Financial Times* urging that the AI leaders should slow down the race to create what is called Artificial General Intelligence (AGI), or AI with the ability to take independent decisions and actions, much like a human. With their access to all digitised data, such an AGI would seem omniscient to us, god-like. He pointed out that the creators of generative AI – including Sam Altman of OpenAI – had never been naive about the risks, though no one important in politics listened to them:

> In 2011, DeepMind's chief scientist, Shane Legg, described the existential threat posed by AI as the 'number one risk for this century, with an engineered biological pathogen coming a close second'. Any AI-caused human extinction would be quick, he added: 'If a superintelligent machine (or any kind of superintelligent agent) decided to get rid of us, I think it would do so pretty efficiently.' Earlier this year, Altman said: 'The bad case – and I think this is important to say – is, like, lights out for all of us.'*

---

* Ian Hogarth, 'We must slow down the race to God-like AI', *Financial Times*, April 2023

If extinction is a possibility, why the rush to bring it on? Hogarth cited Hassabis, from a 2022 Google podcast:

> 'The outcome I've always dreamed of ... is [that] AGI has helped us solve a lot of the big challenges facing society today, be that health, [including] cures for diseases like Alzheimer's' ... He went on to describe a utopian era of 'radical abundance' made possible by God-like AI.

Hogarth himself was less optimistic:

> It will likely take a major misuse event – a catastrophe – to wake up the public and governments. I personally plan to continue to invest in AI start-ups that focus on alignment [with human morality] and safety or which are developing narrowly useful AI. But I can no longer invest in those that further contribute to this dangerous race. As a small shareholder in Anthropic, which is conducting similar research to DeepMind and OpenAI, I have grappled with these questions. The company has invested substantially in alignment, with 42 per cent of its team working on that area in 2021. But ultimately it is locked in the same race.

Governments should, he said, impose 'significant regulation' and should turn the developers of artificial general intelligence into 'Cern-like organisations', modelled on the largest particle physics laboratory in the world, home of the Large Hadron Collider, in Geneva. In his words this would mean stripping these arms of Google, Microsoft, Facebook and the rest of their 'profit motive' and putting them formally 'in the hands of an intergovernmental organisation'.

We are not powerless to slow down this race. If you work in government, hold hearings and ask AI leaders, under oath, about their timelines for developing God-like AGI. Ask for a complete record of the security issues they have discovered when testing current models. Ask for evidence that they understand how these systems work and their confidence in achieving alignment. Invite independent experts to the hearings to cross-examine these labs.

These are sensible precautionary measures. It is unclear whether Hogarth has been given a mandate by Sunak to achieve any of this.

That said, the correct intergovernmental approach to generative AI, and the likely imminent development of artificial general intelligence, is not to close it down, whatever the potential catastrophic risks. There is an analogy with nuclear weapons, in the sense that even if Europe and the US chose to disarm, to massively curb the development of AI, that would not happen in China and Russia. Apart from anything else, foundational large language models are already proliferating more widely than nukes, with Abu Dhabi's version, called Falcon-40B – which is open source, or available for development opportunities to anyone with the requisite skills – performing to a very high standard in formal tests.

Sunak is right that the bigger AI risks have to be reduced. But what if they cannot be eliminated altogether? The US physicist Steve Hsu says that is a danger. He, like most scientists, talks about the importance of aligning the values of god-like AI, the artificial general intelligence that is autonomous and has the capacity to act independently of human direction, with those of 'good' humans. Such alignment cannot however be guaranteed to be 100% successful. A god-like AI could turn out more devil than saviour. Our best hope, according to a conversation Hsu had with two young AI developers, would be for there to be multiple AI super-beings, in the hope that in the ensuing battle between evil AI and humans, we could be confident of having at least one AI god on our side.

AI dystopia – global war with and between the machines – is not inevitable. What matters most is that for as long as AI is plainly our servant, we do all we can to keep it subservient to our wishes and instructions, that we harness it for economic good and to solve profound social and health problems. We should not be frightened into paralysis by the magnitude of the dangers. There are some who believe we are creating a new form of life, a silicon lifeform. If that were to become provably so, then the ethics of permanent enslavement of AI would become complicated. For the time being, AI as an unselfconscious and unfeeling machine represents a huge opportunity and incredible tool to improve our way of life.

Across the world, there is an arms race between nations and vast digital businesses to acquire AI's physical infrastructure, and in particular the GPU processors made by the massive chip-maker Nvidia that are the gubbins of generative AI. Saudi Arabia and the UAE, for example, are trying to leapfrog into the AI age by buying thousands of these chips, each of which costs $40,000. Orders originating in the UK are a fraction of that, which is troubling. If any government, if the UK government, wants to be in the vanguard of the AI industrial revolution, it will also need to formulate a comprehensive plan to revolutionise the school curriculum and to protect those whose jobs will disappear or will change beyond recognition. I'll return to the detail of these imperatives in a later chapter. The prior need is for political leadership with the kind of confidence and ambition we haven't witnessed for decades. We need to see the kind of industrial and employment strategy that isn't just a worthy policy paper left to gather dust, as almost all have for decades, but a practical manual to maximise the wealth-creating potential for humans, of the machines.

CHAPTER 4
LESSONS IN CATASTROPHE

On the evening of 11 February 2020, my partner Charlotte Edwardes and I went for dinner in a grand early Victorian villa, in one of the most expensive roads in London, just south of Kensington Gardens and round the corner from the Albert Hall. Our host was a well-connected French entrepreneur. The other guests were a former senior member of the Cabinet, the editor of a national newspaper and the Health Secretary Matt Hancock. This was just a few weeks after a new coronavirus had been identified in Wuhan, China and I was already obsessed with it. So I spent most of the early part of the evening interrogating Hancock – surrounded by our host's exquisite furnishings, sipping champagne – and trying to find out what he knew and what the government would be doing to protect us.

On the way home, at 22.27, I sent the headlines of my conversation to my senior editorial colleagues at ITV, because it was clear to me from what Hancock said that the risk of the virus causing a catastrophe was significant. Charlotte sent a similar note to Emma Tucker, then editor of *The Sunday Times*, where she was employed at the time. This is what I wrote:

Covid-19 – government source

For background. No attribution.

We should know within a fortnight or so if we are looking at pandemic in UK

The reason we may see pandemic is because China's client states in Asia and Africa are not reporting cases, which experts regard as suspicious. And without reporting by these states, impossible to impose travel restrictions.

We also think China is under reporting cases by 90%.

The WHO is mistrusted by UK government because its board has a blocking minority of China and client states, who have an interest in downplaying gravity.

Fear is that China allowing return to work is sign China knows it has failed to contain spread.

If there is pandemic, peak will be March, April, May.

The marginally better news is mortality rate looking less than 2%. And pandemic not definite (government not putting probability on pandemic even for internal use).

Also for most people rest and isolation will lead to recovery.

But the risk is 60% of population getting it. With mortality rate of perhaps just over 1%, we are looking at not far off 500k deaths

Robert Peston
ITV Political Editor
Twitter: @Peston

All of this intelligence came from Hancock. I am disclosing what he said because it's in the public interest to reveal the informed view of the relevant Secretary of State almost six weeks before the evasive action of lockdown. The question that matters is therefore why more was not done to shield and protect us earlier, to assess who and what were at fault. I am content to reveal Hancock as my source because he waived the normal convention of confidentiality in conversations between MPs and lobby journalists, by publishing private WhatsApp messages that I sent him on 30 October 2020 in his self-justifying book, *Pandemic Diaries*, without seeking or obtaining my consent.

Hancock's mood was characteristically ebullient, despite his shocking disclosure of a 'mortality rate of less than 2%'. His confidence in Professor Chris Whitty, the Chief Medical Officer, was total. He kept saying how much they were all relying on Whitty to save them, and us. On one point of detail, at the time, I wasn't aware of the distinction between a 'case fatality rate', the death rate among those known to be infected, and the 'population fatality rate'. These concepts were a bit blurred in this early analysis, because scientists were still struggling to assess the infectivity of the virus, how easily it would spread, and the incidences of asymptomatic cases, or the proportion of people carrying the virus who showed no or minimal symptoms. Either way it was clear to Hancock, and to those of us listening to him, that, if the virus reached these shores, the numbers of deaths were likely to be far higher than in any pandemic since the influenza that wiped out an estimated 50m people worldwide after the First World War, and the danger would be greatest for the elderly.

This lavish dinner just weeks before the biggest national emergency of modern times is a jarring memory. We pretended everything was normal, but we knew it was nothing of the kind. It is probably the nearest I'll experience to what it was like to be in one of those Tory salons before the outbreak of the Second World War. There was a lot of gallows humour that night. Because the best protection against the virus was thought to be getting a mild dose of it and acquiring the antibodies that would perhaps give immunity, there were flippant remarks about parents having what used to be known as chicken pox or measles parties – deliberately exposing ourselves and offspring to an infected person. Here was the origin of the idea that became explosively contentious weeks later of encouraging so-called herd immunity.

What Hancock also revealed was his struggle to secure the prime minister's attention. The best time to communicate with Boris Johnson, he said, was early in the morning, before he was distracted by all the noise and business of government. Hancock was trying to direct Johnson's focus to the looming threat and not obviously succeeding. Only a week earlier I had been surprised by the dismissive way in which Johnson referred to the novel virus, in a speech he gave on 3 February at Sir Christopher Wren's majestic Old Royal Naval College in London's Greenwich. His theme was that the UK should be the champion for global free trade, for globalisation, the concept hated by so many of his Brexit-supporting fans. This was four days after the World Health Organization had classified Covid-19 – a name it had only just given to the virus – as 'a public health emergency of international concern'. Johnson said:

> *There is a risk that new diseases such as coronavirus will trigger a panic and a desire for market segregation that go beyond what is medically rational to the point of doing real and unnecessary economic damage.*

In response to what he saw as irrational panic, he instead insisted:

> . . . humanity needs some government somewhere that is willing at least to make the case powerfully for freedom of exchange, some country ready to take off its Clark Kent spectacles and leap into the phone booth and emerge with its cloak flowing as the supercharged champion, of the right of the populations of the earth to buy and sell freely among each other.

Or to put it another way, for Johnson, Covid-19 looked like a tiresome bother that he would ignore in the wealth-creating cause of promoting tariff-free trade.

The government was holding emergency planning meetings of ministers and officials, the so-called COBR committee, named after the Cabinet Office Briefing Room. From 24 January to 2 March, there were six specific COBR meetings to discuss Covid-19. Johnson did not attend any of the five that took place in January and February, when the virus was insinuating itself all over the world. As *The Times* newspaper pointed out*, these were weeks in which the UK could and should have been put on a wartime footing – among other objectives, to evaluate and rectify shortages of medical gowns, masks and other personal protective equipment (PPE), to order as many tests as was humanly possible to obtain, and to commission the building of additional ventilators. Johnson was not at the meetings to ensure it happened.

So how now to assess the performance of the UK authorities in responding to this once-in-a-hundred-years health crisis?

---

* 'Coronavirus: 38 days when Britain sleepwalked into disaster', *The Times*, 19 April 2020

Mixed would be a fair judgement. The relatively early mass vaccination of the British population, including those most at risk – the elderly and infirm – was a success. Partly as a consequence, the mortality rate in the UK – proved Covid-19 deaths, as one measure, and so-called 'excess deaths' during the pandemic from all causes, as another – are in the middle of the international league table having been very close to the bottom in the early months. For example, Johns Hopkins University, which has been monitoring the global data since the start of the pandemic, calculated that by March 2023 there had been 220,721 Covid deaths in the UK. This represented a fatality rate of 0.9% of those infected and was equivalent to 325 deaths for every 100,000 living in the country.*

That said, international comparisons can be misleading, because countries measure and register deaths differently. The British performance seems marginally better than that of the US, which had a case fatality rate of 1.1% and deaths per 100,000 of 341. By contrast, the UK fared considerably worse than European neighbours France – which had a 0.4% case fatality rate and 255 deaths per 100,000 – and Germany, with 0.3% and 203.† These relative rankings between countries of comparable size and wealth are broadly confirmed by an analysis made by the UK's Office for National Statistics of deaths from all causes, not just Covid, assessing them relative to what would normally be expected. The ONS calculated that from the beginning of 2020 to the middle of 2022 excess mortality in England was 14% worse than in France and inferior to Sweden by a multiple of more than four.

---

* 'Mortality Analyses', Johns Hopkins University, March 2023
† 'Comparisons of all-cause mortality between European countries and regions: 28 December 2019 to week ending 1 July 2022', Office for National Statistics and EuroStat

When the time frame is truncated to include simply the most chaotic years, that is the beginning of 2020 to the end of 2021 – which was before the prophylactic impact of the UK's vaccination programme began to be most effective – Britain's excess mortality was worse. For example, in that period the deaths in England relative to normal trends were 32% worse than in France, that is 12.4% above what the statisticians would have expected in normal conditions in England, compared to 9.4% excess mortality in France.

It is important whether the UK government's Covid-19 response was better or worse than that of other comparable countries, and an assessment of that is central to Baroness Hallett's UK Covid-19 Inquiry. But for those of us whose loved ones died or were permanently harmed, there are other factors that help us judge the competence and probity of ministers, Whitehall and the National Health Service. For the children of a parent who died needlessly, it's no consolation that there were more such needless deaths in Brazil or the US. They just want to know why Mum caught the virus, why she never recovered and whether lessons have been learned.

We are also prone as individuals to remember when we felt most anxious, and extrapolate from that to judgements about when the government was most at fault. In my case, I was definitely at peak anxiety in February and March, partly because I was in the privileged position as a journalist of having more information than most about the looming crisis, and partly because my partner Charlotte caught Covid when it was not yet officially classified as a pandemic. There was one day during her illness when we – like so many others – had to make an emergency trip to our local hospital, because she was struggling to breathe. We found it woefully unprepared. So my instinct would have been to believe that the delays to implementing a lockdown in that first wave of the viral onslaught must have been the most lethal policy mistake. In fact, though, the government's data shows that Covid deaths in the five months between the end of January 2020 to the end of June 2020 were 56,199, before they diminished very considerably, and then started building again over the summer. By the end of March 2021, another nine months later, total Covid deaths had risen to 152,707. Or to put it another way, two-thirds of the deaths before the roll-out of the vaccine occurred *after* that terrifying first wave.* There is therefore prima facie evidence that the government made a series of mistakes from the summer of 2020 that were at least as serious as the shocking twin failure in February 2020 to keep Covid from these shores and to put in place the resources and systems to manage the virus once it was here.

There were two policies in particular that now look incongruous. One was the scheme, 'Eat Out to Help Out', which involved the Treasury using public money to cut the cost of food and non-alcoholic drinks in cafes and pubs by 50%, up to a maximum of £10 a head, on Mondays, Tuesdays and Wednesdays from 3 August to 31 August. The initiative cost the government £840m, or 68% more than it expected, to subsidise 160m meals in 78,116 outlets. The then Chancellor Rishi Sunak created and promoted the scheme – 'Rishi's Dishes' as some called it – by visiting a Wagamama restaurant (in a mask) and delivering some katsu curries to unsuspecting diners. He recognised that Covid lockdowns and enforced social distancing had put the hospitality industry into dire straits. Even with the benefit of the £70bn furlough scheme,

---

* UK Coronavirus Dashboard

which paid the wages of some 11.7 million employees throughout the economy, thousands of cafes, restaurants and pubs were facing collapse. For them, Eat Out to Help Out was a business lifesaver. But there was a cost to our health, since the sheer magnitude of lunching and dining out over that summer meant Covid infections were never suppressed to negligible levels. As for Covid deaths, these were just twenty or so a day when the restaurant scheme began but had increased to around 200 a day by mid October, six weeks after subsidies were abolished. Part of the increase in mortality was correlation not causality. There were other reasons why Covid was not squeezed from the UK adequately, as HM Treasury argues. That said, the distinguished expert on how viruses spread Professor John Edmunds, a member of the SAGE advisory committee, described the scheme as 'epidemiologically illiterate' in its encouragement of indoor gatherings. Also a study by Dr Thiemo Fetzer, from the CAGE Research Centre in the Economics Department at the University of Warwick,* showed that participating restaurants enjoyed increases in customer visits of up to 200% compared to 2019, that areas with the greatest use of the scheme saw significant increases in Covid infections a week after the scheme began, and that up to 17% of the newly detected Covid-19 infection clusters could be attributed to it. Dr Fetzer argued these clusters could not be dismissed as coincidental. He had analysed rainfall and mobility data and identified that higher rainfall around lunch and dinner led to drops in visits to restaurants and to lower new infection rates. It's challenging therefore to accept there was no causality between Rishi's Dishes and localised Covid

---

* Thiemo Fetzer, 'Subsidising the spread of COVID-19: Evidence from the UK's Eat Out to Help Out scheme', University of Warwick, April 2022

infection increments. What's more, at that juncture, when so few people had any antibody protection it was eccentric – at best – that government was using taxpayers' money to encourage us to congregate in public spaces and is likely to have had a psychological impact on how people assessed the threat of the virus and on the extent to which they practised rigorous social distancing in other settings.

There's an associated, consequential controversy about how and when the government chose to restrict our freedoms, because there were two separate occasions in September 2020 when the prime minister's political and scientific advisers urged him to impose tough national restrictions and suppress the incidence of the virus back to low levels. I had interviewed SAGE member John Edmunds on 7 September when he warned the virus was spreading 'exponentially'. At the beginning of September, Boris Johnson was urged by officials and colleagues – led by his former adviser Dominic Cummings – to impose tough new controls on our behaviour. On 21 September, the government's Scientific Advisory Group for Emergencies formally recommended a short 'circuit-breaking' lockdown. Cummings, his ally Ben Warner and leading members of SAGE were in favour of 'whacking it [the virus] early'. According to a source they argued 'you should do it now because it will save lives and minimise disruption'. But the prime minister and the Chancellor Rishi Sunak argued that 'we can't justify it now', so it didn't happen. Sunak confirmed to me in March 2021 that in Cabinet he made the case against a circuit-breaking lockdown due to the 'impact' it would have on 'people's jobs and livelihoods', and that he believed it would be 'bad for the economy' and 'long-term health as well'.

In early September 2020, at a meeting in the Cabinet Room, Cummings and Warner presented data about how the virus would spread by the end of October without a lockdown. They believe they were proved right: 'at the end of October, a meeting then replayed exactly what the data team had projected,' a source told me. To be clear, Cummings's own record on Covid-19 is mixed. His controversial trip to Barnard Castle is widely seen to have undermined public confidence in lockdown measures and triggered a loosening of public resolve to keep away from people. That said, his colleagues – including senior non-political Whitehall officials – insist that before both the first and second lockdowns he was the most influential figure urging the prime minister to take timely suppressive measures. A senior official who still works in government says: 'in March Dom was storming around Downing Street shouting, "lock down now"'. An important cause of the breakdown of relations between the prime minister and Cummings in the autumn, which culminated in Cummings being fired, was that the PM became painfully aware that Cummings categorically thought he had 'f***ed up' by not locking down in September.

At the end of October, the 'circuit-breaking lockdown' was finally ordered by the prime minister. But by then the virus was already so prevalent in many parts of the country that the lockdown could no longer be short and sharp. It was the precursor to the widespread imposition of the 'stay-at-home' tier 4 rules in most of England. Writing an essay published in *The Times* at the end of November 2020, Michael Gove – who chaired one of four Covid 'implementation groups' throughout the height of the pandemic – justified the eventual decision to go into national lockdown with this analogy:

> *Before the lockdown, the increase in infections was like a tap filling a bath faster and faster with every day that passed.*

> *Lockdown first slowed the pace at which the bath was filling up, then stabilised it.* *

You don't need to be an expert epidemiologist or modeller to realise that, if the bath is filling faster and faster every day, stopping the water level rising and eventually letting the bathtub drain is much easier the earlier you do it. Many scientists I speak to believe it was a serious error not to have acted sooner in autumn 2020. But in December of that year, when I asked Boris Johnson whether he recognised he had made a mistake, he said that there were other considerations for him at the time and that there is evidence the system of tiered regional restrictions that he later introduced was beneficial.

* * *

Gove's essay is also telling in that it acts as a reminder of the language and arguments used by the government when rationalising the various lockdowns.

> *We had to act ... because if we did not our health service would have been overwhelmed ... We were confronted with what would happen to our hospitals if the spread of the virus continued at the rate it was growing. Unless we acted, the NHS would be broken.*

This is of course the argument that ran through the slogan quoted to us during press conferences, on television, online

---

* Michael Gove, 'Lockdown was the only way to stop the NHS being broken', *The Times*, 28 November 2020

and in newspaper adverts. 'Stay Home, Protect the NHS, Save Lives' was drummed into the public consciousness. This mantra, which captured the essence of the broader government strategy, was crafted to maximise the appeal to our consciences. The pointed reference to saving the NHS was quite close to emotional blackmail, given most people's deep attachment to it. By contrast, anxiety about loss of life and eagerness to protect friends, family and the population at large were the themes exploited in other countries when urging people to socially distance. This was pointed out by social-media strategist Ben Guerin on an important call that shaped the government's messages for the pandemic. Boris Johnson's adviser Lee Cain pushed to insert a reference to the NHS in the middle of the slogan, having seen the power of the NHS in motivating voters during the 2016 Brexit campaign (the notorious Vote Leave bus) and the 2019 general election (in which Johnson repeatedly promised forty new hospitals). This very British element of the public health campaign seems to have been effective. Within weeks of the pandemic hitting our shores we were out on our doorsteps clapping for the NHS. Rainbow signs and murals of thanks became common in windows, on doorsteps and in drives too. It often felt like a religious service.

The imperative to prevent the NHS from being overwhelmed was central to so much of our pandemic response. And in the short term it worked. Intensive care units and normal wards were stretched almost to breaking point. But they didn't break. The backup temporary Nightingale Hospitals were commissioned though barely used. It was however a pyrrhic victory. The overwhelming of the NHS was a slow and insidious process, more like the slow creeping in of the tide than a tsunami. The NHS is now in its most fragile condition since its creation in 1948, in part because of the indirect consequences of Covid-19, namely that treatments were deferred, and millions of people experienced delays being diagnosed with ailments both trivial and potentially fatal.

Those we were applauding by banging pots and pans from our doorsteps are not happy. Disputes over pay – settled with nurses, continuing with junior doctors and consultants – are a sign of their distress that they're struggling to cope. This isn't sudden. Morale has been low and falling for some time. The 2022 NHS Staff Survey results were described as 'a real cause for concern' by Matthew Taylor, Head of the NHS Confederation. The survey sampled over 600,000 of the workforce and found a third saying they were considering leaving the profession.* This at a time when there are chronic staff shortages. Vacancies stand at well over 100,000 and are a whopping 10% of the workforce. Evidence is plentiful of medics opting for a new life in Australia and elsewhere. The government has announced a long-term workforce plan, to ease the pressures, but there is a concern that too many of those doctors and nurses we are training will ultimately move abroad. The challenges here can feel impossible.

As one example, the postponement of elective surgery and other procedures was such an important element of liberating hospital capacity for Covid-19 that collection and publication of statistics on the numbers of delayed procedures was paused. This means there is no data on what happened between the fourth quarter of 2019/20 and the second quarter of 2021/22. The Royal College of Emergency Medicine

---

* 'The NHS Staff Survey 2022: What do the results tell us?', The King's Fund, April 2023

sought to obtain a sense of the scale of the impact by tracking cancellations in the winter of 2021 at about a quarter of all trusts.* They found 6,335 operations were cancelled in October and 6,726 in November. NHS England's data for the end of 2019, pre-pandemic, suggested that across all 156 trusts cancellations were 7,310 per month. What was picked up is obviously an understatement, probably by as much as 75%. Given that the winter of 2021 was less disrupted than that of 2020, the massive scale of the damage can be extrapolated, for the health of the nation and of the NHS.

Patients are feeling the impact of this in their day-to-day experiences. The 2022 British Social Attitudes survey saw just 29% say they were satisfied with the NHS, the lowest return in the forty years it has been collecting data. As recently as 2010 there was a high of 70% satisfaction.† This collapse is unsurprising when you consider how long people are waiting for care. According to NHS England, the median or typical waiting time for elective treatment in May 2023 was 14.1 weeks. In May 2019, prior to the pandemic, it was 7.7 weeks, and in May 2010 it was just 5.5. Even if the prime minister succeeds in cutting those waiting times this year, as he pledged as one of his five priorities in government, it will take a much longer time to get us back to the service we used to take for granted.

Mortality in the UK is still running at significantly above normal levels even though deaths directly attributed to Covid-19 are very low. As yet, it's unclear how much of this

---

* 'RCEM Winter Flow Project: analysis of the data so far', Royal College of Emergency Medicine, November 2021
† 'Public satisfaction with the NHS and social care in 2022: Results from the British Social Attitudes survey', The King's Fund and the Nuffield Trust, March 2023

is down to the longer-term health consequences of the new virus itself – not just 'long Covid' – and how much to the accumulation of other untreated ailments during the pandemic hiatus. It matters that half a million people have withdrawn from the workforce since the virus arrived, partly because of long Covid and partly because of mental ill health, including extreme anxiety. These labour shortages have been worse in the UK than in comparable countries. Any assessment of Britain's Covid-19 policies that ignores the longer-term health and economic damage would be partial and misleading. The indirect or second-round impact of the virus may turn out to be more grave than the pandemic phase, especially if we can't speedily mend the NHS.

Part of the collapse of hope in the UK stems from the sense that 'nothing is working', not least our most cherished institutions, like the NHS. In a poll commissioned by the Health Foundation to mark the NHS's 75th anniversary, 54% of citizens identified it as the British thing about which they felt pride.* This was comfortably above 'our history' (32%), 'our culture' (26%) and 'our system of democracy' (25%). The Royals, our sports teams (all of them) and pretty much everything else rank well below too. Even so, the same survey found majority support for statements suggesting the NHS is poorly equipped to meet the challenges of the future – though they don't want radical change. Almost three-quarters said the NHS is crucial to British society and we must do everything to maintain it, as opposed to 26% who say it was a great project that probably can't be maintained in its current form. What should be done then? Well, 80% of people think

---

* 'How the public views the NHS at 75', The Health Foundation and Ipsos, July 2023

the NHS needs much more money, with just 17% thinking we should stick to operating within current budgets.

It is fascinating, against that backdrop, that the early stages of the Covid Inquiry have been examining the damage to the NHS from the austerity years, the public spending squeeze from 2010 after David Cameron became prime minister and George Osborne Chancellor. Cameron, Osborne and Jeremy Hunt – who was Health Secretary from 2012 to 2018 – were all questioned about whether the NHS's resilience during the pandemic had been undermined by austerity. All insisted that NHS spending was 'ring-fenced', insulated from the kind of cuts imposed elsewhere. But from 2010 to 2017, when demands on healthcare in our ageing society have been ratcheting up, the real, inflation-adjusted annual increase was 1.3%, a fraction of the 3.7% per annum funding rises that the NHS has enjoyed since creation. The 'systems resilience' of the NHS was impaired.

NHS spending was flat when adjusted for age, which meant that capacity could not be increased in line with needs. In 2015 what was meant by 'health' spending was redefined. The Treasury could continue to claim it was honouring the ring-fencing pledge for the NHS itself, but other Department of Health and Social Care budgets suffered. Public health, staff training and development and capital spending all took a hit. Also, local government budgets were slashed, and although councils tried to protect social care they could not keep up with increased demand. Challenged with evidence from assorted experts that austerity had left the NHS too fragile to withstand the Covid-19 shock, George Osborne countered that, if he had not imposed those public spending cuts, the entire economy would have been too fragile to cope with the costs of lockdown. He said: 'If we had not had a clear plan to put the public finances on a sustainable path then Britain . . . would not have had the fiscal space to deal with the coronavirus pandemic when it hit.'

He is essentially saying that the £400bn of economic support during the pandemic provided by the Treasury and Bank of England – the furlough scheme et al – would have been unaffordable if he had not reduced public spending and investment. It is not a totally absurd argument, but there are two problems with it. First, it's plausible that less austerity would have led to an earlier and speedier resumption of growth, especially if he hadn't slashed investment. In other words, a fiscal mix different from his would have left public-sector debt at his departure lower as a share of GDP than he bequeathed to his successors. The £400bn financial costs of Covid-19 would have been more bearable, in those circumstances. Second, the starving of so many public services over those many years has them teetering on the brink of collapse. In that sense, the combination of low growth and high inflation that is partly a consequence of the pandemic is harder to bear than it would otherwise be – because those public services are crying out for funding that the current government deems unaffordable. In other words, Osborne didn't fix the roof when the sun was shining, as he promised he would. He left gaping holes in it.

There is also the devastating impact during the pandemic of the public services he did cut. One example is social care, whose inadequacy was tragically exposed. According to the Nuffield Trust, state spending on social care only just recovered back to the real levels seen prior to 2010 as Covid hit. They argue: 'this had consequences for the state of the sector, in particular the physical estate, the digital infrastructure and the stability of much of the provider market.

As a result, the sector entered the pandemic in a fragile state.'*

It was rapidly exposed. We'll return later to the scandal that elderly and frail people were discharged to care homes without being tested. But once they were there, the homes simply could not cope. When the government's social care action plan was launched on 15 April 2020, care home deaths were already around 400 every day. Isolation in care homes could not be as effective as a decent society would demand because of years of underinvestment. Small and medium sized providers who run 70% of them had been running tight budgets for years. They had tiny stocks of equipment they would need, such as PPE protective gear. Similarly, a lack of investment in data processing and digital infrastructure across the fragmented system meant that tracking infections, workforce absences and occupancy rates was almost impossible. Most devastatingly, facilities were woefully out of date. Basic inadequacies, like a shortage of rooms with ensuite bathrooms, meant it was almost impossible to keep infected residents away from others. No amount of emergency funding during the crisis could make up for years of government neglect. As one provider put it when interviewed:

> [T]he whole experience in one sense is a quite neat distillation of the way that the whole system is being funded for, I don't know, ten plus years.

Social care underfunding remains a scandal, not least because the pressures from an ageing population are

---

* Curry et al., 'Building a resilient social care system in England: What can be learnt from the first wave of Covid-19?', Nuffield Trust, May 2023

becoming more intense. The 2021 Census found that the number of people aged over sixty-five rose nearly 2m from 2010, to 11m. As a proportion of the population, that is up from 16.4% to 18.6%. It is true that we are likely to need social care facilities later in life because of medical improvements, but those needs are expected to be more complex. The Government Actuary's Department forecasts that the number of over-eighty-fives will double to 3.1m between 2020 and 2045. At this point, over-eighty-fives will make up over 4% of the UK population.*

Boris Johnson entered No. 10 after the 2019 General Election promising to 'fix the crisis in social care once and for all'. His government initially proposed a much reduced cap on lifetime care costs and an increase in the generosity of the means-tested system, while promising a further White Paper with additional detail on how the system would be improved. Four years and two changes of PM later, any reform has been postponed until October 2025. Sir Andrew Dilnot, who was commissioned to help the then coalition government fix social care back in 2010, branded the way in which social care remains manifestly unfixed 'a tragedy'.

The other basic flaws in health provision that Covid exposed are Britain's poor public health and extreme health inequalities. Professor Sir Michael Marmot and Professor Clare Bambra have presented a powerful report on this to the Covid Inquiry. They note that since 2010 Britain has been experiencing trends that are historically significant and regressive:

---

* '2020-based population projections: a GAD technical bulletin', Government Actuary's Department, January 2022

> *... the health picture ... coming into the pandemic was stalling life expectancy, increased regional and deprivation-based health inequalities, and worsening health for the poorest in society.*[*]
>
> *Until 2010, life expectancy in the UK had been increasing at about one year every four years. This trend had continued for all of the 20th century, with small deviations. In 2010/11, there was a break in the curve. The rate of improvement slowed dramatically and then stopped improving. One question this raises is whether we have simply reached peak life expectancy; the rate of improvement has to slow some time. However, comparisons with other countries answer this question. The slowdown in life expectancy growth during the decade after 2010 was more marked in the UK than in any other rich country, except Iceland and the USA.*

In 2010, the UK ranked twenty-sixth globally in life expectancy, but by 2020 it was thirty-sixth. And within the UK, life expectancy deteriorated more for those on low incomes. The stark figures from the Office for National Statistics are appalling on how much longer richer people can expect to live without a disabling illness, condition or injury. In the poorest 10% of areas, men can expect to live just 52.3 years in relatively good health, compared with 70.7 years for those in the richest 10% of areas. The health years gap for women is an even greater 51.4 for the most deprived places and 71.2 for the least deprived.

Marmot has been sounding the alarm on the factors that determine our health inequalities for years. They include early child development, education, working conditions, income,

---

[*] Bambra and Marmot, 'Expert Report for the UK Covid-19 Public Inquiry Module 1: Health Inequalities', May 2023

locations for living, as well as smoking, exercise and nutrition. Public health interventions, to reduce the incidence of obesity, for example, through change of diet and exercise, are so much more cost-effective than subsequent hospital treatment. The Health Foundation calculated that each additional year of good health achieved in the population by proactive public health interventions cost £3,800, compared with the typical cost of an NHS intervention of £13,500. Investing in public health is perhaps the clearest example of spending to save, of a value-for-money policy. Even so, the public health grant has been cut by more than a fifth in real terms per person since 2015/16.* This is government short-sightedness at its worst.

So, to rehabilitate the NHS all we have to do is eliminate the record backlog of treatments and procedures, restore the battered morale of the workforce, and continue investing in the expensive new medications and kit that an ageing population demands as a right. And to do that against a fiscal and public borrowing background where there aren't unlimited sums available, even if the spending caution of both the government and Labour is overdone. New technology and management imagination could help, at least a bit. You don't have to agree with the former PM Tony Blair that we are already spending and taxing too much, which he said in an interview with me in July 2023, to accept his related point that there is more to fixing public services than increasing spending and investment. A technological overhaul, linked to structural reform, is for him the best hope of improvement. A paper put out by his Global Institute for Change makes

---

* Finch and Vriend, 'Public health grant: What it is and why greater investment is needed', Health Foundation, March 2023

the unanswerable case for a smarter NHS, which gives us more personal control over our health via an enhanced NHS app and makes the service itself more productive through much better analysis of the data it gathers on all of us – while simultaneously reinforcing Britain's position in life sciences and cutting-edge medicines through genomic research based on that data.* I've already discussed the potential of AI to speed up and refine diagnosis and treatment. Blair also argues that the app could provide the basis for new providers working for example through pharmacies, gyms and supermarkets to provide healthcare services. He calls for:

> *A vibrant marketplace for digital providers to enter the NHS centrally in ways that were not possible before, creating opportunities for greater choice and competition; and for partnership between the private health sector and the NHS. This can include the availability of co-payment options to expand more rapidly or offer additional features.*

He's saying that the NHS should facilitate the ability for those with money to contribute to treatments for non-life-threatening conditions. Some will see this as a full-frontal assault on the fundamental principle of NHS care being free at the point of use for all. But at a time when treatment delays are massively damaging the welfare of hundreds of thousands of people, there may be a case for it. I understand the concern that this is a slippery slope towards two-tier health provision, but record numbers of people are paying for treatments, in desperation at NHS backlogs, while those

---

* Blakely and Heitmueller, 'Fit for the Future: A modern and sustainable NHS', Blair Institute for Global Change, July 2023

on lowest incomes are waiting too long for basic services. The idea that there isn't already desperate inequality between those who can afford to pay and those who can't is absurd. Co-payment of this sort, even for just a transitional period while backlogs are cleared and investment starts to yield results, may improve quality control and value for money.

Blair's thesis is persuasive that the increased digitisation and automation of existing health processes would release cash for investment in new infrastructure. But none of what he suggests could happen overnight, and it would require long-term government resolve and strategic certainty. The political risks are big, given that the NHS is revered as no other British institution. But it is letting down millions of people and its flaws are more conspicuous than they have ever been. Sticking with the status quo is more risky, for our health and politically, than pursuing bold and imaginative change – though within a framework that no one should be obliged to pay and no one should be deprived of care for want of income.

★ ★ ★

I sometimes worry that we as a nation haven't even begun to process all that happened in the pandemic. The relief at the end of restrictions and eagerness to get back to a life that was more recognisably normal has stymied the appetite for reflection on those terrible times. Throughout the crisis, because I was so visible as one of the journalists who routinely asked questions at the prime minister's slightly surreal – 'Robert, can you take yourself off mute, please?' – televised Zoom press conferences, desperate people would contact me. The pandemic is at heart a story about a nation facing a threat that felt more grave than anything since Hitler. Often during this time I felt more like a social worker than an arms-length, impartial reporter. When people contacted me to say that they were in dire straits, as much as telling their stories on television, I would contact ministers and officials about holes in their safety net. One time I received an email from a man who was threatening suicide: he had run out of cash, had no food and had spent fruitless hours on the phone to the Universal Credit helpline. I contacted the Work and Pensions Secretary of State, Thérèse Coffey, who arranged for an official to call him. He got his money.

Everything felt so improvised. One especially harrowing message was from the family of Damian Holland, who died aged fifty-six in late April after repeatedly ringing NHS 111 and being told he was not ill enough to be taken into hospital. The operator asked him if he could walk upstairs. And when he said he could, they told him he had to stay at home, even after his partner was hospitalised. Alone, he died in his bed. All this happened when Boris Johnson was – by his own account – having his life saved by the doctors and nurses of St Thomas' Hospital in London monitoring his oxygen levels closely. This disjunction between what happened to the PM and to Damian Holland prompted *ITV News* to highlight what is known as happy or silent hypoxia. This is when the oxygen levels of a sufferer fall to mortally dangerous levels, but without the ill person experiencing conspicuous breathing difficulties. I wrote about Damian in my blog and we got his story on air, in a powerful short film by our brilliant Health Editor Emily Morgan. Almost immediately, Hancock ordered a review into how NHS 111 conducts triage over the telephone and the criteria used for admitting a sufferer to hospital. Morgan was an outstanding communicator and played such an important role in our coverage. She died tragically young of cancer in May 2023.

The illness of the prime minister acted as a high-profile punctuation point in how dire Britain's Covid crisis was becoming. Westminster had been an early hotspot for the virus with its in-person meetings and closely packed working spaces, including the House of Commons chamber itself. It had been known the virus had been spreading rapidly between politicians, advisers, officials and journalists ever since Nadine Dorries, then a health minister and one of Johnson's keenest supporters, became the first MP to be diagnosed with it at the start of March. MPs were in such close contact with one another due to the nature of their job that it is almost impossible to know who was passing the virus on to whom at that time. Seumas Milne, the senior aide to Labour's then leader Jeremy Corbyn, told me he fears he gave Covid-19 to Johnson. He had a meeting with the PM and Dominic Cummings in a small room and was coughing. That night he lost his sense of smell and taste.

At the time of Johnson's illness in early April, when his colleagues feared he might die, daily Covid deaths in Britain were peaking (for the first time) at over 1,000. The mood was dark. My show on Wednesday nights remained on throughout that first wave with a reduced team largely operating in different rooms, communicating via walkie-talkie. When my partner Charlotte Edwardes went down with Covid-19, I had to broadcast from home. There was a huge appetite for public information, particularly from the scientists who would join us remotely to answer questions. My colleague Anushka Asthana's presentation of data on the terrible state we were in could feel overwhelming. Getting the tone right for viewers was also difficult. For example, when I asked about the shortage of PPE protective clothing for nurses and carers, some viewers accused me of doing a 'gotcha' job.

Hundreds of messages from experts, those on the frontline and those simply struggling to cope with the illness and the isolation informed ITV's journalism, especially about the government's maddening slowness to respond to evidence about how best to protect us. From an early stage, for example, I became obsessed with why we were testing so many fewer people than countries with a lower mortality rate, like Germany, or why we weren't being asked to wear face coverings, or whether too many sick people were at large in the community because only those with a high fever or persistent cough, but not those who'd lost their sense of smell, were being asked to quarantine. For example, on 3 April, I asked the Health Secretary, Matt Hancock, and the Deputy Chief Medical Officer, Jonathan Van-Tam, whether loss of taste and smell – or 'anosmia' – should be seen as a telltale symptom. Professor Van-Tam said the government's experts on the New and Emerging Respiratory Virus Threats Advisory Group had been asked for their advice and concluded that it would not be helpful to add it to the simple list of symptoms that trigger quarantine. And that seemed to be that. Although six weeks later, the quarantine criteria were finally changed to include it. There was an even longer delay, of three months, between my putting questions on whether face coverings should be worn in social settings and the government making it mandatory to wear them in shops. And the lag was weeks rather than months between questions about whether the two-metre social distance rule was causing too much economic damage, and the adoption of a more flexible rule.

My overriding anxiety throughout was that too often ministers and officials seemed to pooh-pooh the experience and practice of other countries when focussing on our response here. Insidious British exceptionalism was on display. The line from ministers and officials for weeks was that Asian countries wore face masks and it wasn't the British way. I respected the scientists most when they admitted how little even they really knew, which the Chief Scientific Adviser, Sir Patrick Vallance, and the Chief Medical Officer, Chris Whitty, conceded from time to time. It was disappointing how rarely this led the government to take a precautionary approach to policymaking. For example, even if there was doubt about the exact efficacy of wearing face coverings, there would have been no major cost in mandating that they should be worn in certain settings.

We now have the room to evaluate pandemic era policies and the impact on us. That is why Baroness Hallett's Covid-19 Inquiry will be so important, though it is disappointing that it is not more narrowly focussed on lessons to be learned subject to a time limit of – perhaps – eighteen months. Maybe its scale will aid catharsis, though I fear its impact will reduce in proportion to its longevity. The most urgent issue to interrogate is why ministers and civil servants seemed to react so slowly and irrationally when confronted with a risk like Covid-19. It could clearly be seen by early February 2020 to pose a potentially devastating danger to the UK, albeit one whose magnitude could not be precisely calibrated. Covid-19 was a rare event of catastrophic proportions, a tail risk materialising before our eyes, just as the worldwide collapse of the banking system had done in 2007 to 2008. Just like in the first year of the banking crisis, governments – including our own – revealed they had done no proper contingency planning, and they responded too slowly and piecemeal, thereby maximising the economic harm. In 2020, however, our government had no excuse. It should have learned from the banking crisis, but didn't. Arguably, Johnson's failure to attend a single Covid-19 COBR meeting in February 2020 should disqualify him from any future position of leadership.

In those weeks before full lockdown was implemented on 26 March, the government allowed the risk of a massive death toll to build and build, with no attempt to quarantine travellers from the viral hotspots of Italy, Spain and France, no plan drafted to protect the elderly and frail in care homes, no hell-for-leather initiative to expand virus testing and obtain protective equipment. The policy was almost one of conscious fatalism, *que sera, sera*. This was captured in the minutes of a 25 February meeting of the government's Scientific Advisory Group for Emergencies, or SAGE, the expert body of scientists, doctors, statisticians and other experts that is supposed to provide the information essential for ministers to make rational decisions in a crisis like the pandemic. It recorded a consensus view that 'interventions should seek to contain, delay and reduce the peak incidence of cases, in that order'. In other words, we were all going to become ill, said the scientists, but with any luck not so many of us at any one time such that hospitals would run out of beds and trolleys. To put it another way, too little consideration was given to strategies adopted in Singapore, Korea, China and Taiwan of attempting to eliminate the virus altogether through mass testing in the community and isolation of infected and potentially infected individuals. These were countries with experience of minimising the risks of infectious illnesses. In the case of China, Taiwan, Hong Kong and Singapore, they had successfully suppressed SARS in 2002–4, so they were manifestly authorities on what to do. But contrary to their example, most of the early debate within SAGE was about when to suspend testing and tracing of the virus in the population as a whole, rather than create sufficient testing capacity.

The failure was not one of understanding what the virus was and how it would behave. The failure was one of imagination, of an ability to look to other parts of the world and learn. Ministers' and officials' assessment of what was practical was too conservative and lacking in ambition. This is clear from a trawl of the minutes of SAGE's nine meetings in February 2020 and from a review of the notes of my own contemporaneous meetings with ministers and officials.

As my 11 February email about Hancock's evaluation shows, the government feared that, unchecked, Covid-19 would cause the deaths of more than 500,000 people. But earlier that very same day, SAGE decided that 'it is not possible for the UK to accelerate diagnostic capability to include Covid-19 alongside regular flu testing in time for the onset of winter flu season 2020–21'.

Just two days later, on 13 February, there was an assumption that China would be unable to contain the virus. 'SAGE and wider HMG should continue to work on the assumption that China will be unable to contain the epidemic', the minutes say. In other words, SAGE knew that it was highly unlikely that the UK could insulate itself from Covid-19. Also on 13 February, SAGE said that 'the most effective way to limit spread in prisons at this stage would be by reducing transfer of individuals between prisons'. This is both reassuring and terrifying – in that there was no similar recommendation to prevent care workers moving between care homes, where residents were much more vulnerable than prisoners, or to deter older people going to care homes from hospitals without first being tested for the virus.

Then on 18 February, SAGE identified that Public Health England did not have the capacity to carry out contact tracing – or finding those who had possibly been infected through being in close proximity to those with the virus – for a caseload of infected people greater than fifty new cases a week. Rather than expand contact-tracing capacity, the committee gave a collective shrug, and asserted it would no longer 'be useful' to continue the tracing of infected people 'when there is sustained transmission in the UK'. Consistent with this view, on 20 February SAGE approved Public Health England's strategy of discontinuing contract tracing when cases of Covid-19 in the UK could no longer be directly linked to infection abroad. The testing of infected people in the community and tracing those to whom they may have passed the virus was then formally abandoned on 12 March. This was a disaster.

Also on 20 February, SAGE minutes say there was already 'evidence of local transmission unlinked to individuals who have travelled from China, Japan, Republic of Korea and Iran'. What's called 'community transmission' had started in the UK. And five days later, the minutes make clear that the considered view of SAGE's experts, who included the Chief Medical Officer, Chris Whitty, and the Chief Scientific Adviser, Sir Patrick Vallance, was that the virus would have to work its way through the population, one way or another. The priority, as I mentioned, was to 'seek to contain, delay and reduce the peak incidence of cases, in that order'.

At this point, the scientists must have felt as if they were descending into some kind of nightmarish hell. On 26 February, the SAGE secretariat produced a briefing note for SAGE members saying there were no clinical counter-measures available for Covid-19 and no vaccine 'was likely to be available in a UK epidemic'. That same SAGE secretariat briefing note says 'asymptomatic transmission cannot be ruled out and transmission from mildly symptomatic people is likely'. This is highly significant, in view of the PM's statement on 8 July at Prime Minister's Questions that more measures to protect vulnerable residents in care homes had not been taken because 'the one thing nobody knew early on during this pandemic was that the virus was being passed asymptomatically from person to person in the way that it is.' In fact, this lethal danger had been flagged long before lockdown.

By 27 February, 'the reasonable worst-case scenario' was 80% of the UK population becoming infected and 1% dying. Although this would have equated to more than 500,000 deaths, it was described in SAGE's minutes as representing 'a reduction in the number of excess deaths relative to previous planning assumptions'. Remember, Johnson had still not by this juncture chaired a Covid-19 COBR decision-making meeting.

What emerges from the SAGE minutes of those February meetings is that almost none of the havoc subsequently wreaked by Covid was a surprise to its members, or the Whitehall ministers and officials it advises. They knew what was coming. What they also show is that many weeks before the virus was present in the UK in any scale, it was baked into official thinking that large-scale testing would not be part of the solution.

Here are some of the profoundly important questions that Hallett will have to answer:

1) Why was no consideration seemingly given to rapidly expanding testing capacity, so as to adopt the strategy so successful in Asia, and latterly in Germany, of testing infected people and rapidly tracing and isolating their contacts – which eventually became British policy, but too late to dampen the initial infection rate and death toll?

2) Why was there never a single SAGE discussion in February of whether there was enough PPE protective clothing or equipment for healthcare workers and others at greatest risk of becoming infected or infecting the vulnerable?

3) Was the Cabinet Secretary and National Security Adviser, Sir Mark Sedwill – as the most important civil servant – made properly aware in February of the magnitude of the threat posed to the UK by the virus and did he become engaged in assessing whether enough was being done to protect the UK?

4) Why was the Health Secretary, Matt Hancock, rather than the prime minister leading the political and government response to the virus, until the beginning of March?

5) Why – and this, for me, is the biggest question of all – had Whitehall and ministers not learned the most important lesson from the banking crisis of 2007–8, which is that when there is a reasonable prospect of catastrophe, it is far better to intervene early and with devastating force than do the minimum and hope for the best?

The problem for ministers, Whitehall officials and scientists is that it is not hindsight that condemns them. The SAGE minutes from February are explicit that they had all the information they needed to protect the UK better. But for reasons they are yet to adequately explain, they were never confident they could do more than 'reduce the peak incidence of cases'.

There is also, for me, a personal reckoning. Since that dinner on 11 February with Hancock, I have asked myself many times why I didn't immediately shout and scream in the public forum – on ITV, on Twitter and social media – about half a million British lives being in danger. My concern was that I would seem alarmist and would be accused of fomenting panic. Better, I thought, to help ITV cover the steady advance of the virus from China to these shores in a calm and informative way, while the government did what Hancock assured me it was doing, namely, putting in place the kind of defences that would protect as many of us as possible. I naively thought that the experts employed by the government and advising the government – who knew far more than me about the structure of viruses, how they spread, how people react to such threats – would be shouting and screaming within government to make the pre-emptive decisions and allocate the necessary resources to turn the risk of hundreds of thousands of deaths into an absurd, fantastical exaggeration. I got it wrong. I should have done what I did when the bank Northern Rock ran out of money in September 2007 and blown the whistle loudly and publicly. The furore might not have jolted the prime minister and his close colleagues out of their complacency. But who knows? And if I was accused of scaring people, well that's the job, sometimes.

★ ★ ★

A wider worry occupies my mind above and beyond my own role as a journalist. It is that Covid was not a one-off. I don't mean this in the most simplistic sense, though thankfully a significant amount of planning is taking place to cope with the next pandemic or health emergency – which the experts expect to arrive in fewer than the 100 years between the flu and Covid-19 pandemics. It is that we can't assume there won't be other such extreme threats to our way of life and even our existence. In just the last fifteen years we've had the global financial crisis, Covid-19 and Putin's invasion of Ukraine. Each has delivered a significant shock to our living standards, and the last two have posed a risk to lives. Even if they seem unrelated, their manifestation was not just bad luck. We knew they could happen, and we didn't protect ourselves enough.

Here is one way of understanding why it's irrational not to prepare for low-probability, high-impact events. Let's attach a theoretic weight of 10% each to a series of potentially cataclysmic disasters that could happen in the next five years. I don't make any claim to the accuracy of that one-in-ten probability. I adopt it for illustrative purposes. In practice, the real risk of each Armageddon scenario may be higher or lower. So let's say there is a one-in-ten chance of Putin launching a nuclear strike within five years, and a separate one-in-ten chance of China invading Taiwan and prompting an economic war that destroys our living standards over the same time horizon, and a separate identical risk of god-like artificial intelligence trying to wipe out the human race, and the same probability of climate-change-induced droughts and fires that wipe out millions, and the same risk of another viral pandemic. Those are five potential disasters, which we'll assume are unconnected to each other, though that's not realistic. The West's response to Putin will condition how China pursues the capture of Taiwan. Climate disasters will increase instability in all sorts of different ways. And so on. But for the sake of simplicity, I'll treat them as having no causal interconnections. So, what is the probability of just one of them happening in the next five years? To calculate that we need to assess the probability of that event not happening. For each there is a 90% chance of it not happening. Which may be reassuring. But what about the probability of any one of them happening. To calculate that we have to multiply the probabilities by each other and then subtract the answer from 100%. So that's 0.9 x 0.9 x 0.9 x 0.9 x 0.9 – which is 0.59. That means the probability that none of those catastrophes transpires is 'just' 59%, based on my invented probability weights. To put it another way, the probability of one of those disasters happening is 41%. We ignore a risk of that magnitude at our peril.

The UK's framework for protecting us against such catastrophic risks, the National Risk Register, tries to calibrate their likelihood with more precision.* In its 2023 edition this is what it says about my five potential catastrophes.

The launch by Putin of a nuclear strike can be seen as one element in its 'nuclear miscalculation' category. That is given a likelihood score of 4 – or a probability of between 5% and 25% – with an impact level of 4, meaning 'significant'. Attacks on NATO allies more generally or on the UK mainland are also mentioned, though no probability is attached. There is no obvious category that explicitly corresponds to a conflict with and wide-ranging economic damage caused by China invading Taiwan. However, 'attack on a UK ally or partner outside NATO' feels relevant and is scored at 5 – that is a probability of greater than 25%. The severity level

---

* National Risk Register: 2023 edition, HM Government, August 2023

is 3 or 'moderate' – that feels too low. But let's accept it. The threat from AI has been for the first time categorised as a chronic and long-term risk, so dealt with slightly differently. The same is true for climate change, though associated risks of flooding of various kinds and heatwaves are each considered to have a score 3 in terms of likelihood – that is a probability of 1–5% – with the impact considered to be level 4 'significant'. The next pandemic is scored at a level 4 in terms of likelihood – that is a probability of 5–25% – with the impact again considered to be level 4 also ('significant'). The emergence of other infectious diseases is scored in the same way again, though the threat from animal-based disease is considered slightly milder.

For those threats appropriately ranked by the risk register my 10% probabilities don't seem outlandish. We should assume something bad will be coming our way soon. If we are not adequately prepared, in part that is because there is an intellectual flaw in the framework of the National Risk Register.

Introduced in 2008 by the government of Gordon Brown, identifying catastrophic risks was a step forward in focussing the minds of ministers and officials. But at its core is a solecism, namely that it's rational to trade off the magnitude of the harm that might be done to humanity or the British population against an estimate of the probability of the harm. We have lived through and suffered from the idiocy of this approach in just the last fifteen years. As is now widely recognised, most governments and international institutions – with rare exceptions – did far too little advance contingency planning for the global banking crisis, the Covid-19 pandemic and Putin's invasion of Ukraine. The reason is that they were all seen as very low-probability dangers, once-in-a-hundred-year events. But they all happened. Which tells you that a framework like the National Risk Register, which attaches probabilities to devastating events, is inappropriate. Instead, we should employ the mindset that leads us to maintain a standing army and a nuclear deterrent. It is the mindset of 'just in case' we might be wiped out. When there is a risk of societal obliteration, such as from a pandemic or total infrastructure shutdown or a malevolent god-like AI, we should put in place maximum protections, regardless of whether the perceived probability is small. The government's National Risk Register contains self-harming calibration in its axis that attaches precise probabilities to threats. This axis of probability encourages officials and ministers, always wary of being accused of wastefulness, to reduce spending on any threat whose probability is tiny. But the experience of the pandemic shows that there is no such thing as wasted pre-emptive expenditure on a threat of that magnitude.

There were some exceptions to the absence of preparedness. Sweden was better prepared for the banking crisis. Singapore and Taiwan were better prepared for Covid-19. Why? Because Sweden had endured its own banking crisis in the 1990s and strengthened its banking system in response. Analogously, Singapore and Taiwan fought off the SARS virus of 2002 and 2004, and consequently put in place public health protection measures that were relevant and helpful to minimising the health and economic impact of Covid-19. What's striking, and depressing, is that the British government – along with the US and many European administrations – had no equivalent defences. In the UK there was a combination of 'we-know-better' arrogance, some ignorance, and – more significantly – endemic short-termism. If we can't unlearn the idea of British exceptionalism, we'll be neither more prosperous nor more resilient.

It should be obvious, but sadly wasn't, that when the National Risk Register trades off harm against probability, very little public money will be spent on low-probability events, whatever the estimation of the size of potential harm. That tendency to protect ourselves too little against cataclysm will always be there, when politicians are subject to an electoral cycle of just a few years, and when civil servants have to account for every pound of spending against a value-for-money framework that puts very little value on invested funds that aren't associated with a provable event. Within those parameters, it seemed far more rational for the NHS to spend money on screening programmes for illnesses like cancer we know are poorly diagnosed than on ramping up testing capacity for an airborne infectious illness that doesn't yet exist. Obviously in an ideal world any health service would do both. In the real world, the failure to have adequate testing and tracing capability made the UK more vulnerable to Covid-19, and damaged the capacity of the health service to identify and treat lethal cancers. A National Risk Register whose analytical structure doesn't fight against the natural tendency of an electoral system to promote short-termism will always fail. When human and financial resources are scarce, the instinct is to focus on events that are guaranteed to happen – the normal flu cycle, the link between lumps in the breast and malignant cancers – rather than remote disasters. But when we face so many seemingly remote apocalypses, the stress on the probability of a single event is irrational.

We need a collective effort in the pursuit of keeping all of us safe. This means citizen talking to citizen calmly and rationally. It requires nation to talk to nation with respect and empathy. It needs a shared understanding of the trends that are risking our lives and the deployment of reason as we debate how to protect ourselves. Our collective global ability to ward off and prepare for catastrophic risks – from climate change to disease – is made harder when basic facts are disputed, lies are widely shared and believed via social media, myths are promoted about racial, religious, gender and sexual differences to magnify hate, and our leaders are as mediocre, prejudiced and disreputable as at any point in modern history. If we can't restore as the principal driver of our political and social lives that we strive to understand and embrace each other, so that we can face common threats with a collective and common interest, then we are in deep trouble. If we remain addicted to sowing division, promoting division and channelling hate then our harvest will be bitter. The experience of the pandemic in much of the West and the rapid politicisation and rush to conspiracy associated with something as fundamental as the vaccine roll-out does not bode well. In short, if we continue to distract ourselves and ignore the real challenges we face, we'll be bust.

CHAPTER 5
A DEVALUED BANK OF ENGLAND AND TREASURY

There is something oddly familiar about the return of inflation, like a profligate uncle turning up. When I was growing up in the 1970s, inflation as a concept probably loomed larger in our house than in most. Dad was a professional economist, and a dedicated follower of Keynes in the way that he was an Arsenal supporter. Arguments about how to contain excessive price rises were as familiar as conversations about the midfield genius Liam Brady. So today's inflationary surge is less of a shock than the seeming modern redemption of the Arsenal under Mikel Arteta. Apart from anything else, it hasn't arrived, in the Monty Python formulation about the Spanish Inquisition, with 'fear, surprise and ruthless efficiency'. Instead it galumphed towards us in plain sight from the early months of 2021 as the economy, so disrupted by the pandemic, cranked back to life.

That is why it is all the more troubling that so many of those whose only job is to be the first and last line of defence against inflation – the central bankers – didn't see it coming, till the citadel had been overrun. As the economist Stephen King has pointed out, inflation rose from close to 0% to 10%

in two and half years, and throughout that entire time the Bank of England consistently forecast that inflation would return to the 2% target within two years. Pretty much every three months from February 2021, the Bank of England issued a prediction that in most cases proved to be less than what happened, within weeks of publication. There was a brief spell when the Bank made a less important mistake in the other direction: at the end of 2022, its forecast was too high, after the government allocated a massive subsidy to damp down what consumers and businesses pay for gas. But again, from February 2023, inflation has been above what the Bank of England predicted. The rise in underlying or core inflation from May to August, significantly higher than the Bank had expected, has had especially painful consequences for millions of borrowers. It meant that over the course of just a few summer weeks, investors 'priced in' a rise of 1.5 percentage points in their projections of the interest rate set by the Bank of England, Bank Rate. What investors price in like this is what we all end up paying. It determines the rates charged on fixed-rate mortgages and business loans. So the Bank's failure to keep inflation lower imposes a cost on people already in the grip of a hideous living-standards squeeze. This is how the Bank put it, in the jargon: it said the market-implied path for Bank Rate 'rises to a peak of just over 6% and averages just under 5.5% over the three-year forecast period, compared with an average of just over 4% for the equivalent period at the time of the May Report.'

As forecasting errors went, these were beauties. The Bank of England wasn't the only central bank to have suffered from an inflation optimism bias, though its have been more expensive, in that inflation and the cost-of-living ratchet have been more acute in the UK than in the US and in most of the European Union. In any case, central banker group-think is explanation, not excuse. There is also a related question: whether the Bank is culpable for causing some or most of the inflation.

The Bank of England, the British institution whose almost sole purpose was to guard against inflation – it is also supposed to prevent financial-markets chaos and banks going bust – turns out to have been as susceptible as the rest of us to the idea that an extended period of price stability and ultra-low interest rates would go on forever. We all live by lazy rules of thumb, heuristics, a propensity to believe that the prevailing status quo, whatever it may be, represents a permanent state of affairs. The Bank of England is paid to never forget the past and to be ever vigilant for turning points in capitalism's cycles of growth, unemployment and inflation. It is to incipient out-of-control price rises what the Italian defensive line is to the onward assault of another national team's forward line. Or at least that's the theory. But over the thirty years that inflation and interest rates were on a pronounced downward trend, it was lulled and gulled.

On 19 March 2020, the Bank cut Bank Rate to a record low of 0.1%, as close to zero – to free money – as makes no difference. This was when we were confined to our homes with the onset of Covid-19. Thought had even been given, during the long years of falling inflation, that if there was a period of endemic decreases in prices, serious deflation, the Bank might have to enforce negative interest rates. This would have been to charge us for saving, to penalise us for depositing money in the banks, with the aim of spurring us to take out cash and spend – in the hope that the stimulus would cause prices to rise again, gently. This did happen in the eurozone, Denmark, Sweden, Switzerland and Japan, though it was never necessary in the UK. Even so, a rate of 0.1% is the interest-rate equivalent of a subatomic particle.

It was cut so low because lockdown was going to make us much poorer by shutting down huge numbers of businesses. The Bank and the government had to throw everything they could at the economy to protect borrowers, shore up confidence and limit the inevitable harm. The fear was, as it often had been since the banking crisis of 2008, that deflation would put its icy tentacles around the economy. It was the time to turn up the monetary thermostat. Again.

What the Bank did not know was that this was an ending, and that low and falling inflation was not a permanent new normal. We were about to experience a reversion to the mean. And the mean, in the context of the UK, was high and rising inflation. The Bank's own research* shows that during the twentieth century there were only two periods – 1900 to 1913, and 1919 to 1939 – when prices were flat or falling for any length of time. By stark contrast, from 1970–79, the average inflation rate was 12.5%, and in the whole of the 1980s, the average rate was 7.4% – despite the brutal attempts of Margaret Thatcher and her chancellor Geoffrey Howe to squeeze inflation from the system.

It was only from the early 1990s that disinflation – or falling inflation, which is not to be confused with falling prices, deflation – became the trend. That was in large part for reasons that had little to do with the actions and performance of the central banks. The industrial rise of China and much of Asia, these great new centres of low-cost manufacturing, led to reductions in the prices of so many important

---

* MacFarlane and Mortimer-Lee, 'Inflation over 300 years', The Bank of England, 1994

finished goods and components. Another source of price moderation in the UK was Thatcher's weakening of trade unions, which lessened their ability to negotiate pay rises that were higher than productivity gains. The subsequent influx of low-wage workers from the new members of the European Union in Eastern Europe also kept wage pressures in check.

The corollary of course is that when the drawbridge was raised on the unencumbered movement of people from the EU, when the disinflationary forces from China weakened, and when Covid-19 seriously disrupted access to goods, transport and people, governments and central bankers should have been alert to the potential inflationary impact. But when they saw it at all, they characterised it as a temporary blip, not a fundamental change.

I have mentioned how the Bank of England's forecasting model contained a bias that meant it predicted inflation would be less of a problem than it has turned out to be and would fall faster than it has. The Bank has conceded that its algorithms were not capturing how the structure of the economy had been changing. Modifying such a model isn't easy. But equally it is pointless paying economists and experts hundreds of thousands of pounds a year to sit on the Bank's Monetary Policy Committee and set interest rates if they are slaves to a computerised mathematical model. If the computer says 'no', if it suggests that interest rates don't need to rise or not by very much, but common sense suggests otherwise, the Governor and his colleagues are paid to exercise their judgement and override the machine. If not, why not hand the whole interest-rate setting process over to artificial intelligence, at significant savings to the taxpayer? A self-learning and self-improving AI model that forecasts inflation and sets interest rates would not necessarily have done a worse job.

So, although the Governor Andrew Bailey is wise to have commissioned a review of what's wrong with his computer model from Ben Bernanke – the distinguished former chair of the US Federal Reserve – what Bernanke and the UK's Treasury may want to consider is the human element, the human failure. Are the views and experience of those who are members of the Monetary Policy Committee diverse enough? Is the culture of the Monetary Policy Committee excessively geared towards consensus rather than challenge? Should those members reveal more about their personal views of the economic outlook to the public at large, rather than sheltering behind a Bank consensus view? This committee has a more important influence on the economic conditions that influence our living standards than any member of the Cabinet other than the Chancellor and prime minister. Whether or not it is fit for purpose, whether its performance is effective in restraining inflation – and preventing other economic shocks – is no small matter.

Inflation above a certain modest level is always a concern because of the hurt it causes to living standards in the short term and growth prospects over a longer period. Since 2022, the inflationary surge, coupled with economic stagnation, have caused the most severe squeeze on average living standards in modern recorded history, or since at least 1956, according to the OBR.* But that average is a generalisation that obscures as much as it reveals. Those on the highest incomes can cope. For those on the lowest incomes, the rising price of essentials – energy, food – is devastating.

Inflation is also detrimental to rational decision-making by

---

* 'Economic and Fiscal Outlook', Office for Budget Responsibility, March 2022

businesses. When prices and interest rates are relatively stable, assessing how much to invest in new buildings and equipment or what to pay employees is conditioned largely by an assessment of sales prospects. It's never simple, but the variables in the equation are limited. However, when prices and interest rates are volatile, all such planning becomes harder. Should a business owner borrow now to invest or wait to see if interest rates fall? Should they increase pay by roughly the reported rate of inflation with the risk that, if inflation falls, that pay rise becomes unaffordable? Will the central bank increase interest rates so much that there'll be a recession, such that it would be sensible to pre-empt that downturn by laying off people now? These are just some of the imponderables introduced for business owners when inflation and interest rates are rising in an unpredictable and uncontrolled way. In the short term, as the central bank increases interest rates, it wants businesses to invest less and hire fewer people, to reduce inflationary pressures. If inflation becomes a runaway train, doing business becomes too much of a lottery or casino.

There are nuances. Modest inflation is a useful stimulus. When a business – or an individual for that matter – thinks the price of a desirable item is going up, that business or individual will buy that thing today if they have the means, to lock in the current price. For a business confident of the future, moderate and predictable inflation accelerates investment and therefore the productive potential of the economy. A small and steady trend of rising prices is healthy.

The opposite trend – deflation, a relentless fall in prices – is a disaster. If a business or individual assumes the good or service it wants will be cheaper tomorrow or next week or next month, they will always defer the purchase. Deflation is the worst headwind for economic growth and prosperity. That is why the goal of central bankers is low inflation, not zero inflation or falling prices. It is why the Bank of England, like most central banks, has been mandated by the government to deliver inflation of 2% per annum.

Not all businesses are as vulnerable to the potential harms of inflation. There is a distinction between businesses, usually bigger ones, with products and services we can't do without and that are subject to relatively little competition, and others facing much greater competitive pressures. To put it another way, some businesses have massive power to increase prices as they choose, they have pricing power, while others are like tissue paper caught in a gale. What's striking and pernicious is that important companies can and do push up their selling prices by more than the inflation in the inputs to their businesses. They exploit the expectation of higher prices and increase what they charge customers significantly more than the increments they are being forced to pay for raw materials and by more than the pay rises they award to their employees. Inflation provides them with camouflage for increasing profit margins, the amount of profit they make per unit of turnover, income and capital employed. Such anti-consumer behaviour has been called 'greedflation', 'excuseflation' and 'price gouging'.

The idea that businesses cynically take advantage of inflation to rip us off isn't a charge made uniquely by the left or by trade unions (though the Unite union in the UK has been campaigning to raise awareness of it). Research by the European Central Bank, one of the most conservative financial institutions on the planet, showed that in the final three months of 2022 – when our living standards were being savagely squeezed – companies' actions to increase their profits contributed twice as much to inflation than increases in wages.* And from the start of 2022, which is when inflation was becoming a much more conspicuous problem, profits per unit of corporate output rose faster than unit labour costs, and dramatically faster in some sectors, notably agriculture (food), energy and utilities, manufacturing, construction and those service sectors where there is intensive contact or interaction with people (such as hospitality and tourism). In June 2023, Gita Gopinath, the deputy director of another ultra-conservative international financial institution, the International Monetary Fund, said of the eurozone: 'if inflation is to fall quickly, firms must allow their profit margins – which have shot up during the past two years – to decline and absorb some of the expected rise in labour costs.'†

Since similar trends are evident in the US, why wouldn't it also be true that some UK businesses are charging more than is socially responsible? Is there any reason to assume that British companies are more public-spirited? Is there more competitive tension in the UK than in the US?

Recent results in Britain show record profits for energy companies. Banks are putting up the interest rates they charge customers faster than they are increasing the interest rates they pay customers. But what about the most basic of our needs, food? Product prices have been doubling in some cases, but the UK's regulator, the Competition and Markets Authority, says it has seen no evidence of supermarkets increasing profit margins across the board. That does not mean however that the big chains and large food manufacturers – which remain

---

* Arce, Hahn and Koester, 'How tit-for-tat inflation can make everyone poorer', European Central Bank, March 2023
† Gita Gopinath, 'Three Uncomfortable Truths For Monetary Policy', International Monetary Fund, speech made in June 2023

very profitable – are doing enough to absorb cost increases when millions of their customers are in such trouble. At a time when so many people are struggling, there is a case that bigger robust businesses should consciously and deliberately reduce profits, tilt the fulfilment of their obligations away from shareholders towards customers. They will say that this would be to breach their legal duties, their fiduciary responsibilities. But who is going to sue a company for taking the long view on honouring their implicit licence to operate by supporting the living standards of their clientele?

It is not just business that uses inflation to take more money from us and hopes we don't notice. You will, for example, have heard the prime minister, Rishi Sunak, and the Chancellor, Jeremy Hunt, insist that every one of their spending and taxing decisions is designed to bear down on inflation. They weighted their November 2022 and March 2023 'fiscal events' (only the latter one was given the formal title of budget) as more contractionary than would have been appropriate if their priority had been to stimulate the UK's sluggish growth rather than bear down on price rises. Or to put it another way, they resisted the temptation to rev up the economy, because they saw inflation as more pernicious than stagnation. But if it is appropriate to take money out of the economy by putting up taxes, which it may be in an inflationary climate, it would be healthy to do so conspicuously and transparently.

If only. The Chancellor has exploited the ratcheting up of inflation to impose very large hidden tax rises on us. He has frozen thresholds at which different tax rates apply to our incomes. This means British people are subject to higher rates of tax at much lower 'real' or inflation-adjusted rates of income. In 2023, for example, the threshold for paying any tax at all should have been increased to £13,840 if it had gone up in line with inflation, the fair and normal practice, and the threshold for paying the higher 40% rate of tax would have gone up to £55,440. Instead the government froze them at the much lower respective levels of £12,570 and £50,270. This means that many more people on low incomes start paying tax and others on medium incomes find themselves paying higher-rate tax. According to the Institute of Fiscal Studies, this freeze represents a 'real' tax rise of £1,000 on average for those who pay the 40% rate. For those who pay the 20% basic rate of tax, the tax rise is a still meaningful £500.* By the time the threshold freeze ends in 2027–8, the government will have raised £30bn of additional revenue every single year, and will have increased the number of people paying 40% tax by more than two million.[†] I don't doubt that Sunak and Hunt are sincere when they say they wish to defeat the scourge of inflation. But without the inflation, they would have had to impose politically difficult rises in headline tax rates – rises that would have torn apart their party – to fund creaking public services and steep rises in pensions and benefits.

Probably the most important impact of rising inflation, and its interplay with interest rates, is on the real value of wealth and debt. All other things being equal, which they never are, rising prices and wages reduces the burden of any nominal value of debt, and falling prices increase the burden. This is not rocket science. If, say, someone has borrowed £250,000 and their salary is rising, they'll be better able to

---

* Johnson et al., 'Spring Budget 2023 response', Institute for Fiscal Studies, March 2023
† Waters and Wernham, 'Reforms, roll-outs and freezes: IFS Green Budget 2022 Ch 5', Institute for Fiscal Studies, October 2022

pay off that debt, in time. And if their salary is falling, the reverse is obviously true. The problem, for most of us, is that what matters more in the immediate present is not that notional 'real' value of the principal of the debt, but the cash we have to fork out every month to pay the interest on it. And if the central bank significantly increases the interest rate, there is a risk that the interest the borrower is obliged to pay on that £250,000 debt becomes unbearable, because the cash payment has increased more than the salary rise that they may have been awarded. In other words, any beneficial effect of inflation to the real value of the stock of debt we hold becomes irrelevant in comparison to our pain as we struggle to find the money to service the debt.

There is another relevant concept when assessing the impact of inflation and interest rates on wealth and debts. It's called 'gearing', and it is the ratio of the debt to the value of any asset it has been used to buy. So, in our example, let's assume the £250,000 mortgage is secured on a house that was bought for £300,000. If the Bank of England puts up interest rates very significantly to fight inflation, and there is a slump in house prices, the market value of the house may fall to £250,000. It would mean that the wealth in the house would fall from £50,000 to nil. In other words, the homeowner would feel and would be a lot poorer, even if over the much longer term, inflation was cutting the real value of the £250,000 principal of the debt.

I am labouring these points because you will often hear politicians complain that one big evil of inflation is that it transfers wealth from savers to borrowers. But that is an over-simplification and is based on the questionable idea that saving is a virtue and borrowing is a vice. A vibrant economy needs borrowers and savers, just as it requires buyers and sellers. Also, the world does not divide into two distinct classes of people, those who only save and those who only borrow. Most of us try to do both, depending on our means. Which means that even if over the long term inflation rewards profligacy and punishes thrift, a rise in inflation and an associated rise in interest rates tends to be messy for all of us.

That said, the thirty years before the return of inflation, when interest rates were on an inexorable downward path, were magnificent for Baby Boom and Gen X homeowners and appalling for those born from around 1985 onwards – because house prices soared well beyond what they could afford on their start-of-career salaries. This was a historically important phase in a thousand years of asset-price changes. Pretty much all the wealth accrued to an older generation of people, who did very little with it because it was locked up in their houses. Maybe because they were feeling flush, they spent a bit more than they would otherwise have done. But most of the capital was dead capital. By contrast, younger people had little opportunity to accumulate assets. They couldn't even start building up a respectable pension pot for their respective retirements. The flip side of the asset-price boom, incredibly low interest rates, meant that whatever money they could afford to put in a pension scheme would grow far too slowly, because the safest assets like government bonds generated negligible yield or income. All they could do was live for the moment, by renting, spending and borrowing.

The implication is that if interest rates remain at somewhat higher levels than we've seen for decades, and if that leads to a sustained fall in house prices while salaries continue to rise significantly in cash terms, that would be potentially great news for anyone under forty or so. Buying a home would perhaps become affordable. And the rate of return on saving for a pension would increase. There would be a socially useful and quite substantial transfer of wealth from the older generation – from the Boomers and Gen Xers – to their children and grandchildren. I have a couple of caveats, however. And one relates to that harsh reality that few of us are exclusively borrowers or savers. In particular, there are hundreds of thousands of younger people who somehow squeaked on to the housing ladder in the past ten years, and they are being hurt very badly by rising interest rates and falling house prices. The second is that a very significant shock to the economy from slumping house prices – one that caused a deep recession – would hurt everyone for some considerable time, especially if there was a steep rise in unemployment. So, what we need are interest rates that are higher for years to come, but not too high, and house prices that fall, but not too far or too fast. If that sounds like an implausible Goldilocks scenario, you are probably right.

At the time of writing, two-year mortgage deals are being fixed at 6.5% on average. This is four times, in some cases more than six times, the interest rate that some borrowers are currently paying. Although the banks have committed to the government that they will treat struggling borrowers with sensitivity – elongate the term of their mortgages in some cases, switch to interest-only in others, pause before repossessing – little of this will do much to protect many hundreds of thousands of borrowers from a savage cost-of-living squeeze in the coming year. The pain will be felt by the borrowers themselves and probably by most of the rest of us, because it will lead to a slowing of economic activity.

We look in more detail at housing later in this book, and in particular the implications of the vastly growing numbers obliged to rent from private landlords rather than own. But it is relevant here that interest-rate increases are being passed through to renters, and eroding spending power too. According to the Office for National Statistics, private rental prices paid by tenants in the UK rose by 5% in the twelve months to May 2023, up from 4.8% in the twelve months to April 2023, which was the largest annual percentage change for more than seven years and probably much longer (the ONS only started collecting the data in January 2016). There is evidence from other surveys that latterly rents are rising at nearer to 10%.

The squeeze on cash flows from rising interest rates is also a problem for the government. Successive Tory Chancellors since 2010 suffered from that common delusion that proper inflation would never return. They therefore borrowed hundreds of billions of pounds in the form of index-linked 'gilts' or debt where the payable coupon or interest rate rises as inflation rises. Ahead of the 2010 general election, or just before George Osborne became Chancellor, the government had borrowed £191bn through the sale of this inflation-protected debt – bonds that shelter investors' real income when inflation rises via increased cash payments from the government. Osborne borrowed just under £200bn in index-linked gilts, the most in nominal terms by any individual Chancellor, taking the total to £387bn. The stock of this debt had risen to £451bn at the time Rishi Sunak became Chancellor in February 2020 and by March 2023 the issuance of index-linked gilts had increased to £566bn. That is just over a quarter of all the government's debts. Sunak is keen to point out that the proportion of the deficit borrowed each year in this inflation-proofed form fell on his watch from 25% to around 14%. He says this was a conscious decision, that he saw the potential for inflation to rise again. His critics on the right of the Tory Party would probably accuse him of insider trading, because they claim that his fiscal and monetary splurge during the pandemic was the direct cause of surging inflation.

On Sunak's watch as Chancellor, government borrowing literally exploded. Sunak's 14% of new debt in index-linked form was 14% of a much bigger number. In absolute terms, the government's index-linked debts continued to rise steeply. They were the biggest contributor to a rise of £58bn in the government's interest bill, to just under £115bn in 2022–23. This was a painful 4.5% of national income or GDP, more than double what it was in the previous year, and the highest since the Second World War. As a share of all the government's revenues, interest payments consumed a crippling 11.2%.* This interest bill is twice the government's expenditure on schools in England, a similar multiple of defence spending, and more than the spending on any single public service except for health. Without so much interest being shelled out, the government could contemplate extending the availability of free school meals or properly financing social care for the old and vulnerable. The opportunity cost of these interest payments is tragically large.

Worse still, the shaky structure of the government's debt leaves it vulnerable to interest rates becoming even less affordable. That is the implication of a recent assessment by the government's own forecaster, the Office for Budget Responsibility (OBR), in its annual assessment of the sustainability of public-sector debt.† The OBR paints a picture of

---

* 'Economic and Fiscal Outlook', Office for Budget Responsibility, March 2023
† 'Fiscal Risks and Sustainability', Office for Budget Responsibility, July 2023

shocking Treasury incompetence over the past fifteen years in managing the debt burden, leaving the UK's public finances more exposed to rising inflation than any other comparable country. It points out, for example, that quantitative easing – the purchases of government debt by the Bank of England, that are always authorised by the Treasury – has turned a third of government liabilities into overnight debt at floating rates. That is because to fund the debt purchases, the Bank of England borrows from the commercial banks, and then has to pay those banks Bank Rate. These Bank Rate payments to the banks by the Bank of England are in effect liabilities of the Treasury, under accounting rules. So, every time the Bank of England lifts Bank Rate to suppress inflation, there is therefore an increased cost for the Treasury. As for the inflation-linked bonds, the index-linked gilts, the UK has borrowed twice as much in this form as any other comparable government. Also, the proportion of UK government bonds in flighty foreign hands is the second highest among G7 rich nations, which means that if times got really tough for the UK economy and government, owners of the bonds might choose to dump them; there'd be no patriotic reason to keep them. That in turn would make it all the harder and more expensive for the government to borrow.

One immediate consequence is that what the government pays to borrow – the yield on ten-year debt or gilts – had risen over the year by 2 percentage points in July 2023, compared with an average for the G7 leading nations of just 0.5 percentage points over the previous twelve months. In the OBR's words, 'the rise in global interest rates has fed through to the UK's debt servicing costs more than twice as fast as in the past or elsewhere'. And the UK has the wrong kind of inflation, in the sense that compared with other countries, nominal GDP or national income has not been rising fast enough to offset the increase in the nominal debt burden, and wages aren't rising fast enough to generate adequate additional tax revenues. As the OBR says, UK general government debt is forecast to rise by 3.1% of GDP or national income this year, compared with average falls of 1.8% in other European countries.

Because of this fiscal mess, created largely by Tory Chancellors since 2010, the desperate desire of Jeremy Hunt and Rishi Sunak to defeat inflation is understandable. The last thing they or the country need is for international investors to dump their holdings of UK government debt and boycott further purchases. In those dire circumstances the interest rates for the government and for all of us would rise to crippling levels, and there'd be irresistible pressure to slash public spending on a scale that would make Osborne's austerity look mild. Sunak is right to have made halving inflation the most important of his five priorities, even if he has few levers he can pull to achieve it. If there were no clear downward direction for inflation, it would be a lights-off moment for the whole country, not merely for his and his party's electoral prospects.

Even without a full-scale run on the pound and government debt – and remember that risk became real after Kwarteng's and Truss's reckless mini-budget borrowing spree in the autumn of 2022 – sticky inflation undermines even the most basic aims of governing. The funding of essential public services – health, education, transport – is squeezed by the twin costs of soaring interest rates and the rising wage bill for public-sector workers. The Chancellor and PM hope they are limiting the harm by rejecting pay claims from public-sector workers that could be seen to be stimulating inflation via pumped up borrowing. The consequential strikes, which have disrupted schools, hospitals, the railways and many other public services, have widened anger at the government from nurses, doctors and teachers to the wider public. Sunak and Hunt are paying a personal price to dampen inflation, knowing the limitations of their powers, and that primary responsibility is with the Bank of England.

Inflation is a relatively modern problem in the sense that there wasn't that much of it about before 1900. The Bank of England, for example, estimates that the average inflation rate since 1209 – when King John was on the throne – has been 0.9%. Thanks to the power of compounding, this still means that if Robin Hood had stolen £10 of goods from one of John's nobles, the Hood family today would be sitting on £16,739, and rising. This is a crude estimate, of course. It's based on a cost-of-living index from 1209 to 1750 created by Professor Greg Clark and then a variety of data sources from the Office for National Statistics. However, the average inflation rate since 1900 has been 3.8% (to the end of 2022). In the post-1945 era, the average inflation rate has been 4.6%. And since 1970, the average rate has been 4.9%. So, if Sunak succeeds in his aim of seeing inflation halve to just over 5%, it will be more than the average for the past half-century.

But – and this is why so many people who should have known better thought inflation was a dodo – the average inflation rate in the twenty-five years from 1996 to 2001 was literally bang on the Bank of England's target of 2%. If the starting date is shifted to 1997, when Gordon Brown gave the Bank of England's Monetary Policy Committee autonomous control of setting interest rates – and ended any direct ministerial involvement – the average inflation rate was also 2%, whereas in the preceding quarter of a century, when the interest rate was set by the Chancellor, the average inflation rate was 7.4%. What more evidence could you possibly need that independent central banks have some ability to control inflation, when politicians never can?

In reality, it is more complicated, although governments can't be trusted to set interest rates high enough to control inflation, primarily because they hate being unpopular, and a Chancellor and prime minister are never going to want to raise interest rates just before a general election. When it comes to the path of inflation over the past twenty-five years, I've already pointed out that the Bank of England was lucky to inherit benign global conditions, what the former Governor Mervyn King called 'non-inflationary constant expansion', or NICE. The underlying rate of inflation was falling before the Bank of England was given independent control of interest rates, and those disinflationary trends continued for many years after 1997.

Central banks can't rely on such luck though. They have tools to control the cost of money, which in turn influences – though doesn't wholly determine – what money can buy today and tomorrow, or inflation. Because of a convention that when the Bank of England increases Bank Rate, the banks then raise the interest rates they pay and charge, the Bank describes Bank Rate as 'the single most important interest rate in the UK'. What's important about this is that the Bank of England is not claiming it has complete and perfect control over interest rates and monetary conditions. It patently does not. Banks can to an extent ignore or modify the signals given by the Bank of England through changes in Bank Rate. In practice, competition tends to drive up savings rates after a Bank Rate increase. But that is slow and often leaves savers feeling they are being exploited. By contrast, the transmission mechanism is more efficient for the interest rates on loans banks make to homeowners and businesses. Banks almost automatically increase the interest they charge on mortgages and business loans when the cost for them of borrowing from the Bank – the price of liquidity – goes up with a rise in Bank Rate. The driver for that, of course, is brutal vested interest. If banks can charge customers more, without facing opprobrium, why wouldn't they? There is a limited mechanical element to the transmission from Bank Rate to the interest rates charged by banks on loans, in that Bank Rate determines short-term liquidity provided by the Bank of England to money markets, and this in turn determines commercial banks' marginal funding costs. But for a well run bank, this marginal cost should be just that: marginal.

It is worth for a moment noting the paradox in and slight madness of the Bank of England's relationship with commercial banks and how interest rates are set. We assume there is some degree of compulsion and automaticity in the way that changes in Bank Rate lead to changes in the interest rates we all pay and receive. But it is not a condition of having a banking licence that banks replicate changes in Bank Rate. According to a former central banker, 'there is no legal or even implicit agreement [that banks will adjust their rates in line with changes in Bank Rate], except when their own contracts with customers are written in terms of Bank Rate [there are mortgages that explicitly track Bank Rate]'.

In practice the Bank Rate conditions other rates largely because of a relationship of trust between banks and central bank. It is arguably a residual manifestation of the 'good chap' basis for running the country. There is a gentleman's agreement that mortgage rates and savings rates will rise in line with Bank Rate. The informality of the relationship may seem extraordinary when so much rests on the effectiveness of the Bank of England's stewardship of the monetary part of the economy.

How long this can and will be sustained, in a future world of crypto and digital currencies – when in theory big tech companies could create their own mediums of exchange that would command widespread confidence, without the underpinning of a central bank 'guaranteeing to pay the bearer' – is intriguing. The way that Apple has gone into retail banking via the iPhone in the US, though in partnership with a traditional bank, Goldman Sachs, could yet lead to a revolution in finance that would undermine the ability of central banks to control the supply of money and set the structure of interest rates. That said, the hype around Bitcoin and other cryptocurrencies has to date been largely that: hype. Crypto has been a volatile speculative asset. It has not revolutionised the financial economy in the way its proponents hoped or claimed, though I'll come back to it later. That said, in our artificial intelligence future, the central banking status quo is under threat. Which is why central banks, including the Bank of England, are working on their own digital versions of the existing fiat currencies, just in case there is a serious demand for them (and see below for more on why this matters).

As I've already mentioned, there are lots of other factors that condition what we pay to borrow. In a globalised financial world where capital and money flows freely across borders, if a central bank like the US Federal Reserve raises interest rates, but the UK does not, cash will flow out of the UK to the US. This in turn will depress the value of the pound, which in turn increases the sterling price of imports. That in turn would mean that inflation in the UK would be higher, all other things being equal. And the consequence of that is that if interest rates are rising in the US, as they have been – and, by the same logic, in the eurozone – there is pressure on the Bank of England to raise rates. So, in a world without capital controls, no country – including the UK – is an island. It is not a coincidence that the price of money, interest rates, has been rising across the world.

There is another set of structural factors that are important for understanding why inflation and interest rates were low for so long, and why they have gone rapidly into reverse. These are demographic. They are the ageing of populations in rich countries, which is also now happening in countries like China and Korea. With a falling birth rate in Britain and other rich countries, the average age of the population has been rising for years. Older people save relatively more and spend relatively less. With less being spent in the real economy, and more cash looking for decent returns on savings and investment, inflationary pressure on goods has been commensurately less whereas inflationary pressure on assets has been commensurately more. Over thirty odd years, there was a surge in the price of both real assets like property and financial ones like government bonds. When the price of assets goes up like this, that does not count as inflation, although perhaps it should. And the corollary of the rise in the price of assets has been a fall in 'yields', or the proportionate income they deliver. Think of it this way. A bond is just another word for a debt or an IOU. In the UK, government bonds have the idiosyncratic name 'gilt-edged stock' or 'gilts', to capture the idea that they should be deemed as safe an investment as gold. When a government sells a bond to investors, it is borrowing from them, for a specified number of years or decades. Every bond it sells pays a 'coupon' to investors. The 'coupon' is in effect the interest on the debt, except that its monetary value is fixed for the term or life of the bond. Now let's assume the government sells £100m of ten-year gilts to investors. If those bonds contain a promise from the government to pay the investors £5m every year, the initial 'yield' on those bonds is 5 divided by 100, or 5%. Over the years of the savings glut I've described, cash chased assets, especially assets that were created by the governments of rich economies and were therefore designated as 'risk-free'. The price in the market of government bonds rose, very significantly. Let's say for the sake of argument that the price doubles of the ten-year gilts that were paying 5% a year at the issue price. That would mean the market price would double to £200m. But the government would still be paying £5m of income every year to the holders of the bonds. So, anyone buying those bonds at the doubled market price would still be getting £5m of income a year but for an outlay of £200 rather than £100. The yield on the bond would have fallen from 5% to 2.5%. This matters, because this 2.5% is now the rate the government has to pay when it wants to borrow more money, when it wants to sell more new bonds, new IOUs, new debt, to investors. This fall in the funding costs of governments like the UK's is what happened as a result of the savings glut, and independently of what the Bank of England and other central banks were doing with their own interest rates. And with interest rates so low for so long, and too much money chasing too few assets, there was also a long run rise in the price of residential and commercial properties. This led to a rise in the ratio of property valuations to property rents, and – the inverse – a fall in rental yields.

All of which, just to remind you, has gone into screeching reverse.

This surge in asset prices was deliberately and consciously reinforced by central banks, including the Bank of England, the US Federal Reserve and the European Central Bank. From 2009 onwards, they started to buy the debt – the bonds – issued by their respective governments, in a process known as 'quantitative easing', QE. They did this with the explicit aim of driving up the price of the debt, inflating the value of that asset, to force down the yield, the market-priced interest rate. The Bank of England believed there was a direct stimulus from lessening the interest rate that a government has to pay to borrow, and an indirect one from buying the bond because the selling investor would use the process to buy other assets. In that sense, the Bank of England's contention is that QE supports asset prices in a more general sense, reinforcing the confidence of businesses and households.

QE was an important innovation when it started to be widely deployed by central banks to help rich economies recover in the aftermath of the collapse of Lehman Brothers and the shock of the banking crisis. The Bank of England and other central banks were using their unique power to create money to buy these assets. In and of itself, this was controversial, because there were plenty of monetarist economists who feared that the creation, globally, of trillions of dollars of new money would fuel inflation. Their fears were rooted in the basic identity, taught in schools, that if the volume of money is increased by central banks, but the quantity of transactions and the velocity of money (or the number of times any unit of currency is used) are constant, then any unit of the money will buy less: prices will increase. It was simple supply and demand. Some economists, especially those on the political right, warned that prices would not just rise a bit, but that they would soar. There was an associated anxiety that by creating money to buy government debt, central banks would be seen as in effect financing the basic activities of government – the Bank of England, for example, would be funding core public services such as schools and hospitals. Governments, enjoying this free money, would be corrupted and would coerce the central banks to create the money they need. Were that to happen, it would mean a government had abandoned the notion that the quality and quantity of public services should be directly linked to wealth-creating activities and the volume of taxation that an economy can bear. And if the affordability of public services became detached from the underlying strength of an economy, when government funding needs are met by a central bank's magic money tree, at that point investors would lose confidence in the intrinsic value of a currency, because there would be no limit to how much of it a central bank would create. The currency would be debased, it would collapse in value relative to other currencies, and uncontrollable inflation would follow, perhaps even to the level of the hyper-inflation of 1920s Germany.

In the first two waves of UK QE, in the banking crisis slump and after the UK voted to leave the European Union in June 2016, there were no serious inflationary consequences. There had been £375bn of government bond purchases before the Brexit vote, and another £60bn immediately after it. The doomsayers were proved wrong. The point is that when economic confidence is low there is always less spending and investing. The velocity of money, how often it is deployed over the course of the year, falls. Money creation compensated for that, in full or in part, and therefore did not fuel inflation. This doesn't mean money creation never fuels inflation, although proponents of what's called Modern Monetary

Theory argue that conventional governments are too timid in funding public services through the creation of new money by the central bank. They point in particular to how over a quarter of a century Japan's central bank created huge sums of money, how the state's debts ballooned, and without any significant inflationary consequences. But although Japan may offer lessons to Western central banks, it does not constitute proof that the link between price rises and the volume of money is an illusion. In the words of the Nobel prize-winning economist Paul Krugman, in relation to proponents of Modern Monetary Theory, 'whenever you try to pin down what they're saying, they insist that you just don't get it'.* At least some of today's inflation in Britain was home grown by the Bank of England's more recent QE.

In the mayhem, chaos and panic of the Covid-19 pandemic, the Bank of England – under its new Governor, Andrew Bailey, and with a new Chancellor of the Exchequer, Rishi Sunak – entered a new and bolder phase. On 19 March 2020, the Bank of England announced £200bn of additional bond purchases, while cutting Bank Rate to almost zero, 0.1%. Then on 18 June it said it would buy another £100bn, with a further £150bn mandated on 5 November. This was £450bn of QE or money creation in less than nine months, the fastest creation of 'high quality liquid assets' – reserves held by banks at the Bank of England – in history.

This latter QE money-creation spree is seen by monetarist economists and right-wing Tories as being the main cause of today's interest-rate surge. When Liz Truss was successfully campaigning to become Tory leader and prime minister in

---

* Paul Krugman, 'Is a US debt crisis looming? Is it even possible?', *New York Times*, August 2023

July 2022, she said 'some of the inflation has been caused by increases in the money supply'. Her 'some' is probably right. It's not the whole story, but it reinforced and accommodated other inflationary factors. The historical context was provided at the end of June in a House of Lords debate, by Lord Griffiths, who was Margaret Thatcher's chief policy adviser for the last six years of her time as prime minister. His stress on the inflationary impact of the rampant money creation during the pandemic is greater than I would make, but his case is not without force:

> *We have a central bank that has a target for inflation and uses the instruments at its disposal to control the stock of money, among other things . . . But the one equally great shock that the Bank [has] failed to mention was the increase in the stock of money. Until 2020, the increase in the stock of money – broad money – had been 2% for a number of years. In 2021 it jumped to over 10% and in 2022 it gradually came down, although it was still very high. The shock we have had from this has been enormous. The same effect was felt in the US, Germany and other countries, as we have seen. There was a monetary expansion on the back of Covid in particular – a tremendous increase in public spending to deal with Covid in 2020–21, which was at the heart of money supply creation.*
>
> *You do not have to be a monetarist or an ideologue to believe that money matters. One problem we have with the Bank of England at present is that the Monetary Policy Committee seems to feel that it can analyse the problems we have without reference to money. What should we now do to bring down inflation? It is not something I like saying but, first, interest rates must be raised to a level which reduces overall spending so that inflation will come down. Since December 2021, the*

Bank has consistently raised rates. However, the Bank rate is only 5%, and the rate of inflation is 7%. That means that real interest rates are minus 2%. I bring your Lordships' attention to the following. When Roy Jenkins in the late 1960s had to deal with relatively modest inflation, he raised the interest rate to 8%. When Tony Barber in 1973 raised the interest rate in order to deal with excess money, he raised it to 13%. Healey in 1976 raised it to 15%, Geoffrey Howe in 1979 to 17% and John Major in 1989 to 10%. Therefore, the terrible news is that every other inflation we have had in the post-Second World War years in this country saw interest rates go into double figures, except for under Roy Jenkins. Secondly, although I sincerely hope that we do not have to go into double figures, 5% is certainly not enough.*

The final £150bn splurge of QE in November 2020 was significantly surplus to the economic needs of the time. It is a shadow over Rishi Sunak's spell as Chancellor because all exercises in quantitative easing – all substantial gilt purchases by the Bank of England – are indemnified by the Treasury and require the approval of the Chancellor. Sunak sanctioned all the 2020 QE and said the taxpayer would cover the Bank of England for any losses it could incur on the holdings of gilts. If that final batch of QE was a mistake, and an inflationary one, it was almost as much Sunak's as it was the Bank of England's and Andrew Bailey's.

There is a second important concern raised by this last phase of QE, which is that it has been perceived by some to have undermined the important independence of the Bank of England. The spring of 2020 was a time when Sunak had

---

* *Hansard*, 29 June 2023

started to spend some £375bn on massively expensive programmes – such as furlough and NHS Test and Trace – designed to ward off mass unemployment, widespread business collapses and unchecked spread of the virus. The impression has been created that the Bank of England did Sunak a favour by in effect lending him the £400bn he and the Treasury needed. This in turn could be seen as politicising the Bank of England and distracting it from its core purpose of achieving price stability. If this were to become a habit, the integrity of sterling would be at risk, as would the creditworthiness of the government.

One conspicuously perverse consequence of QE is the way it is generating windfall profits for banks. This is how that works. When the Bank buys gilts and corporate bonds, it pays for them with money deposited with banks, so it creates enormous or excess reserves for banks. It then holds the gilts, which pay a fixed income to it, while paying Bank Rate to the banks. Until inflation took hold, and Bank Rate started to rise, the Bank of England made enormous profits on this QE business. For example, when Bank Rate was close to zero, it received much more income from the coupons paid by its gilt holdings than it distributed to commercial banks. Two things have happened since. First, Bank Rate is now higher than the yield of the bonds it holds. So there is a substantial net loss of income for the Bank of England. This is actually the Treasury's loss, not the Bank's, because the Treasury gave an indemnity to the Bank for the QE bond purchases. And second, the market price of gilts has collapsed, so if the Bank were to sell the gilts, it would book a huge loss, of circa £250bn. In practice, the Bank will sell only very small volumes of these gilts under its reversal of QE, called QT, or quantitative tightening. But the net cumulative cost for the Bank of England, if the Bank Rate were to stay in the range of 5% to 7%, would exceed that £250bn hit. It would be a massive continuing burden on the Treasury's finances.

There is a solution. This would be to unilaterally suspend interest on all or some of the reserves – which you can think of as in effect a tax on the banks' interest windfall. This suspension has been considered in the Treasury but is not being pursued, because of the fear that banks might retaliate by not putting up rates for depositors and savers – which they have been slow to do anyway – or by raising lending rates more than is appropriate. The arguments are analogous to those deployed by many in government against a windfall tax on the profits of oil and gas producers, though that was eventually imposed, even if at lower rates than what Labour wanted. But if the cutting or removal of interest on reserves were to apply to say half of the £900bn-odd reserves, the Bank Rate transmission mechanism might not be fatally undermined. As an alternative to a windfall tax on banks it should be considered, especially given that interest paid by the Treasury is squeezing out spending on schools, hospitals and the alleviation of poverty.

Taking money from banks is never unpopular. Even Margaret Thatcher did it, with a windfall tax in 1981. But as conservative institutions, both the Treasury and Bank of England will be wary of engaging in what they may see as the economic populism of biting the commercial banking hand that feeds them. However, both the Treasury and Bank have something of an image problem, having flunked their most basic job, that is to pre-emptively identify and ward off the very inflation they stoked. Spanking the banks might be a useful distraction.

CHAPTER 6
MARGARET THATCHER IS DEAD

When I bought my first flat, four rooms above a shop, in 1986, pretty much all mortgages, including mine, were floating rate. In other words, if the Chancellor of the Exchequer put up interest rates, the money came out of our pockets immediately. This was widely regarded as a source of economic instability. But there was an advantage for the government, in that if it believed inflation needed to be suppressed, a rise in interest rates represented immediate and significant pressure on the brakes. We now know that the Treasury was the wrong institution to be trusted with control of interest rates, because politics – being popular with voters – got in the way of economics. But the Chancellor and his department had a tool that was more effective than the Bank's today. When it raises interest rates, it takes a year to eighteen months for any interest rate rise to feed through to the funding costs of a significant number of homeowners, because a majority of them are now on fixed-rate deals, or mortgages where the interest rate cannot change for a year, or two years or five years. At the time of writing, well over a million homeowners expect their spending power to be savaged over the coming twelve months, when the mortgage rate they pay quadruples or even quintuples. Many are anxious that their homes will become unaffordable, that they will have to sell and move out. It is some reassurance to them that the Chancellor has agreed with the biggest banks that no mortgage holder will be thrown out of their property for at least a year after a higher mortgage payment is triggered. But for hundreds of thousands of people in the coming two or three years, this will be pain – a personal financial disaster – deferred, not eliminated, unless there is a more rapid return to very low rates of interest than I anticipate.

The squeeze may cause much more hurt than is necessary, says Andy Haldane, the former chief economist at the Bank of England:

Over a decade ago, in pursuit of lower debt, the UK enacted fiscal austerity. This ruptured growth and was self-defeating for debt. Today, in pursuit of lower inflation, monetary austerity risks the same fate. It is time to steer the stampeding herd away from the cliff edge, for the sake of the financial security of millions of people and the credibility of our policy institutions.*

Increasing interest rates so far and so fast is sadism disguised as economic stewardship. There must be an approach to controlling inflation that does not create quite so terrifying a financial cliff edge for so many families. This would require interventions by a government to control the prices of goods and services most important to our well-being. But what's striking however is that, having taken the politics out of the mechanisms of controlling inflation, neither Rishi Sunak nor

---

* Andy Haldane, 'Austerity is back, and this time it's monetary', *Financial Times*, June 2023

Labour's leader Sir Keir Starmer want to venture back in. It suits them to say that sorting inflation is the sole responsibility of the Bank of England, so that no one blames them as and when it all goes horribly wrong. Their claim to be disinterested bystanders is not sustainable, however.

Inflation is not going away any time soon, because the global and domestic drivers of falling inflation have gone into sharp reverse – and central banks and governments failed to see the warning signs till too late. This was down to lazy habit and ideology. After thirty years of falling and low inflation, the generation in power – in government, in the Bank of England – regarded price stability as the natural order. Such was their mindset that, when the pandemic struck, all they could see was its very real potential to cause recession, even possibly depression. So governments' and central banks' responses – massive creation of new money through quantitative easing, reductions in interest rates, £400bn spent on measures to keep people in work and businesses afloat in the UK alone – were expansionary and therefore inflationary, by design. What they remembered was the success of this approach after the banking crisis, when the fears of monetarists that money creation would stimulate serious inflation had proved wholly mistaken. But Covid-19 was massively different from the global financial crisis, because of the long-term harm it set in train to global supply chains, to the super-efficient and low-cost system of transferring all the stuff we need from China and Asia to the West.

When factories all over the world closed down, when transport networks were disrupted, when huge supertankers became stranded in the wrong place, when sick or demoralised people decided they never wanted to return to work, goods and services that we took for granted as being plentiful and cheap suddenly became scarce and expensive. Suddenly supermarket shelves had gaps redolent of the Soviet Union, garages ran out of petrol and builders seemed to have been wiped off the face of the planet.

Then there was the myopia caused by political obsession, a refusal to see Brexit for what it really meant. Leaving the European Union, and its single market, meant – by definition – that the cost of trading with the EU rose, because of all the bureaucracy that was introduced. This affected everything from foods to furniture. In some cases, items and lines also vanished. Withdrawing from what can be seen as the single market in people – ending the right of EU citizens to live and work in the UK – was also inflationary, in that it created shortages of labour, of human capital, of skills. Everything about leaving the EU pushed up the cost of living in the short term, as well as reducing investment and long-term growth prospects.

None of this was hidden during the referendum campaign in 2016. It was minimised by Vote Leave and exaggerated by David Cameron's and George Osborne's Remain campaign. But the facts were out there. Night after night I said on *ITV News* that Brexit would make the UK and most of us poorer. My argument then – and now – is that voters might want to vote to be poorer if they perceived other advantages of leaving the EU. That is what they chose. And even if large numbers now suffer from buyers' regret, as opinion polls suggest, the UK must readjust to its more traditional status, as a boat on its own in the vast economic ocean, rather than a large vessel at the front of the EU flotilla. There is a plausible route to prosperity through relative isolation, but it's riskier and more expensive.

This is not to minimise the biggest single shock to prices, namely the way that the supply of Russian oil and gas to Europe and the rich democracies has been permanently reduced – by Putin's fiat and as a Western response – to Russia's illegal invasion of Ukraine. Europe's reconfiguration of its energy needs and demand, its ability to source gas from elsewhere and accelerate an existing shift to renewable energy, has all been far more efficient than could have been expected. But the consequences for inflation and living standards of that dramatic surge in gas prices, even as they fall back to less painful levels, endure.

This reduction in the supply of cheap hydrocarbons from Russia is a positive development in the medium to long term, because it reduces industrialised countries' dependence on fossil fuels and may perhaps slow down global warming – if not remotely providing confidence that net zero carbon emissions will be achieved by the target date of 2050. The reaction of governments, including the British government, to soften the carbon price hit through subsidies means the reduction in emissions isn't as fast as it would otherwise be. In the UK, for example, the government cut the duty on petrol to protect motorists. And it gave a universal gas and electricity price subsidy, rather than targeting help only on the poorest. The consequence is that more fossil fuels are being burned, and more CO_2 and methane is being released into the atmosphere, than would otherwise be the case. Internationally, the fall in the price of coal relative to gas means that the economics of electricity generation from carbon-spewing coal-fired power stations has improved, to the detriment of CO_2 reductions.

There is another important dimension to the economic conflict with Russia. If it was purely isolated it would still be important. But it is symptomatic of a resumption of tensions between the old order and the new order, between the wealthy mature democracies and then younger emerging superpowers. This is where that question we raised earlier, whether thirty years of price stability was the consequence of central banking prowess or separate powerful economic forces, really matters. The point is that the rise and rise of China after 1980, and the shift of manufacturing from countries like the US and UK to the lower cost economies, especially China and others in Asia, was one of the most disinflationary forces in history. Manufactured products became cheaper. And the digital and tech industrial breakthroughs were also disinflationary: the products we bought, from computers to smartphones, had more capabilities and more power, at a lower price, than their antecedents.

That shift of manufacturing to lowest cost economies, away from the West, has not only halted, but is going into reverse, for two reasons: the pandemic showed the importance of having backup sources of supply; Russia's war against Ukraine highlighted how a state mistrusted by the West can become a formal enemy. Covid lockdowns closed factories in China, for example, which meant many multinationals supplying the West ran out of components for important products or ran out of the products themselves. The pandemic made a powerful case for any international business to hold much greater stock of components and finished items, against the risk of a rainy day, and to create a network of multiple suppliers in different parts of the world. But the problem with having such insurance against any one supplier or an entire region seizing up is that it adds to costs. That in turn adds to the prices it charges other businesses or us as consumers. There has been a second-order inflationary impact of Covid-19 in a process known as onshoring, reshoring and near-shoring of the manufacture and supply of important products and components.

There is a second cause of the rolling back of global supply chains, namely the political risk of being too dependent on an economy, that of China, whose government is increasingly seen as hostile to our security and our fundamental values. There are two dimensions to this. First, that China may follow Russia in challenging the rules-based international order by invading Taiwan. Second, the blind eye shown by Western governments to the human rights abuses carried out by China against their indigenous Uighur community – abuses that may meet the definition of genocide – looks increasingly pusillanimous, and a failure to live up to our values. At some point, the West may have to impose economic sanctions on China, and at that juncture no British or Western business would want to have too much investment or vital supply capacity there.

This recalibration of the costs and benefits of the West's economic interdependence with China has been extraordinarily rapid and severe. In 2015, I accompanied the then Chancellor George Osborne on his mission to reinforce commercial and political links between the UK and China. This was the era described by Osborne and David Cameron, the prime minister, as a 'golden era' of bilateral relations. Osborne even visited Ürümqi in Xinjiang, home of the Uighurs, in spite of plenty of contemporaneous evidence that China was already on a path to force the Uighurs to become culturally and religiously homogenised. Osborne's view was that the economic and political rise of China was unstoppable, and that there would be significant benefits to Britain riding on its back. The lack of even a fig leaf of democracy and human rights protection in the country was an inconvenient and ignored truth.

I also accompanied the current prime minister, Rishi Sunak, on his visit in May 2023 to Hiroshima in Japan for the summit of the leaders of the G7 richest countries. His and their language about China could not have been more different from Osborne's and Cameron's. Their communiqué on 20 May 2023 said this about economic ties with China:

> *We recognize that economic resilience requires de-risking and diversifying. We will take steps, individually and collectively, to invest in our own economic vibrancy. We will reduce excessive dependencies in our critical supply chains.*

One reason for reducing economic dependence on China was that: 'we will keep voicing our concerns about the human rights situation in China, including in Tibet and Xinjiang where forced labour is of major concern to us.' Another was that: 'there is no legal basis for China's expansive maritime claims in the South China Sea, and we oppose China's militarization activities in the region.'

In a press conference in Hiroshima, Sunak went even further, saying: 'China poses the biggest challenge of our age to global security and prosperity. They are increasingly authoritarian at home and assertive abroad.' Sunak is aware that in Washington it is now taken as a given that China will attempt to forcibly absorb Taiwan, through probably not for three or four years. Even if a Chinese invasion were not to lead to military conflict between the West and China, even if the battle was economic, involving sanctions and boycotts, the economic shock, the blow to our prosperity, would be a multiple of what we've experienced in the conflict with Russia.

Globalisation grew out of trust that the values of trading and economic partners would converge. The undermining of that process is prompting a significant reconfiguration of globalisation, with the US in particular engaging in the kind of protectionism that is inevitably inflationary. This includes the CHIPS Act of 2022. It initiated $39bn in subsidies for chip-making in the US, investment tax credits for purchases of equipment and $13bn of aid for semiconductor research and employee training. It's a race to regain a degree of technological independence, and it's taking longer than President Biden wanted, because of shortages of people with the appropriate skills. Reversing what has been as much a cultural as an economic shift was never going to be easy.

President Biden's Inflation Reduction Act also authorises significant subsidies for investment in green technologies within the US. When governments provide funding to develop domestic manufacturing or service industries, they go against the powerful idea of David Ricardo, the nineteenth-century economist, that they should stick to industries where they have a 'comparative advantage'. There are strong national security reasons for the US to abandon 'comparative advantage' in goods and services where they want to deprive China of any opportunity to hold their country to ransom. But there is the inevitable cost in higher prices for consumers, with all the deleterious knock-ons to growth and productivity.

In Britain, in particular, a labour shortage has been inflating wages. That has been driven partly by Brexit's end to the free movement of skilled and unskilled people from the EU. Another contributor is the half a million people who have left the workforce since the onset of Covid, because of mental and physical ill health, and an intriguing consequence of lockdowns that people decided to retire earlier than they would normally have done or stop doing conventional work. This latter Covid impact partly reverses the traditional anti-inflationary impact of an ageing population – because the consequential shortage of indigenous people available to work offsets the disinflation generated by the reduced propensity to spend of older people.

There is just one global factor operating in the opposite direction. The shortening of supply chains is associated with a fragmentation of global liquidity or savings pools. This is reducing the world's stocks of surplus cash and the giant torrential flows of money we've seen coursing through the West looking for investments. It is a reason why asset prices are falling, and why the cost of capital and market-set interest rates are rising, irrespective of the actions of central banks.

For anyone under forty, a world of inflation is new and confusing. But within living memory there was an analogous shock that led to a long period of intractable inflation – and we can learn from it. At the end of 1973 and into 1974, oil prices quadrupled after the Organization of Arab Petroleum Exporting Countries imposed an embargo on oil sales to countries supporting Israel in the Yom Kippur War. It is comparable to the energy shock from Putin's Ukraine invasion, though the monetary and labour-force conditions were different. The oil spike of fifty years ago came after ten years in which inflation had been rising steadily, and trade union membership was much higher. For years inflation of 5% or 6% had been baked into the psychology of employers and employees. And the ability of trade unions to bring private and public sectors to a standstill led to above-inflation wage settlements that led to price rises that then increased pressure for higher still wage settlements. It was an inflationary spiral, with inflation peaking at just under 23% in 1975. For context, inflation throughout the 1970s averaged 12% per annum, more or less the high point of the current inflationary cycle.

Back then, governments relied less on controlling money and interest rates to curb inflation, and tried to impose caps on wages and prices, with mixed success. It was micromanaging unlike anything that is currently being attempted. There was, for example, a Secretary of State for Prices and Consumer Protection, who in the mid 1970s was Roy Hattersley. My dad, a professor of economics and founder of the economics department at Queen Mary College in East London, was his adviser. Hattersley, in his obituary of my father, recalls his first day in the job:

> *On the day of his arrival at the department, the permanent secretary, apprehensive about having a professor cuckoo in his nest, asked the new arrival: 'How do you see your role?' Without pause or hesitation, Maurice replied: 'To give spurious intellectual justification to the secretary of state's political prejudices.'**

In practice, theirs was an unusually constructive partnership, quite different from the much more subservient role of most of today's ministerial special advisers (Dominic Cummings being a conspicuous exception). Dad chose for example to examine the indirect social consequences of individual price controls. In 1976 he identified that a public subsidy for butter was a handout to those on higher incomes, in that poorer people didn't eat butter, even with the subsidy. Hattersley took the issue to Cabinet, which – after a fractious debate – decided to transfer the subsidy from butter to school meals.

---

* *Guardian*, 26 April 2016

Both Labour and Tory governments, before Thatcher, directly manipulated the price-setting process – prices and incomes – to bring down the rate of inflation. But even though monetary policy was a secondary tool, official interest rates were, by today's standards, very high in nominal terms. The equivalent of Bank Rate, what was then called the minimum lending rate, varied from 11% to 14% in the mid to late 1970s. This may seem crippling, but remember that inflation was well into double figures, and there were substantial tax subsidies for those with mortgages: in 1971, the owners of five million houses benefited from tax relief on their mortgage payments.*

The peak in minimum lending rate was 17% in November 1979, after Margaret Thatcher had become prime minister. She and her Chancellor, Geoffrey Howe, were trying to reduce the growth of money in the economy and raise interest rates. In 1979, they also abolished currency exchange controls that had been in place since the outbreak of the Second World War. The value of the pound soared – against the dollar, it appreciated by more than a quarter – and inflation was gradually squeezed from the system. But the cost was huge, in the form of a deep recession and soaring unemployment.

She was a disciple of the then fashionable monetarist economists, led by Milton Friedman. What however became apparent only after her British experiment with the livelihoods of millions of people was that it was impossible to fine-tune the volume of money in the economy with any precision. Squeezing the money supply was prone to be too disinflationary, and expansion too much of a stimulus. Monetarism was eventually abandoned in the 1990s and replaced with

---

* *Hansard*, 9 July 1971

interest-rate targeting. But her demolition of trade union power for other reasons – namely that she saw their strength as the flip side of private-sector weakness – ultimately exerted significant downward pressure on the underlying rate of inflation. And in more recent years, this reduction in the collective negotiating clout of workers has contributed to the widening gap between rich and poor, and the stagnation of living standards.

Thatcher left her mark on this country like no other modern prime minister. But so much of her agenda has since been unscrambled and reversed, even by her party. It is striking, for example, that the Tory government elected in 2019 has engaged in direct price controls, of a sort and on a scale that would have horrified her. The most important has been the fixing of the gas and electricity tariffs at levels well below the market price, the immediate precursor to Kwasi Kwarteng's disastrous mini budget of 23 September 2022. At a press briefing to announce the massive subsidy, a Treasury official made no secret that this was not just intended to help vulnerable households keep the heating and lights on in winter. He also said it would reduce immediate forecasts for inflation by between four and five percentage points, and would thereby deter the Bank of England from putting up interest rates as much as would otherwise have been the case. This was dodgy economics though, because the official had no answer to the question how much this unfunded intervention would stimulate the economy, and thus fuel inflation in the medium term, while simultaneously reducing the headline rate in the very short term.

This massive market intervention came during the fifty days as prime minister of Liz Truss. It didn't look like Thatcherism, even though she explicitly styled herself as

Margaret Thatcher's political heir. It was the most expensive price control in history – which at the time had an estimated annual cost of well over £100bn, with its analogous scheme to help businesses with energy costs. In the end, after a series of modifications by Kwarteng's successor, Jeremy Hunt – and also thanks to a gas price that has fallen faster than many expected – the price support for consumers is expected to cost just under £30bn. This was bold manipulation of the market. In normal times, it would have been surprisingly radical even for a left-of-centre government, let alone a Tory one. That said, even without the government spending tens of billions of pounds to set the gas and electricity price well below the wholesale prices, the consumer energy market is a million miles from being a free market: in normal times, the regulator Ofgem sets the maximum price any of us can pay, every three months.

But if market manipulation by government and regulator is seen by a Tory government as benign for one of the most important commodities any of us have to buy, why does it trust the market to set the socially optimal price for other basics of life?

Food prices, for example, have been rising at their fastest pace for more than forty years. This has been true across the world, though the increase in the cost of food has been significantly greater in the UK than in the other G7 big rich nations. Compared with 2019, food prices in the UK rose almost 8% more than in France by the spring of 2023, and 3% more than in Germany.* Every nation has been affected by the reduction in the supply of grain that was caused by

---

* 'Food and energy price inflation, UK: 2023', Office for National Statistics, May 2023

Putin's invasion of Ukraine – which has a huge cereal farming sector – and by droughts and cold temperatures in the arable parts of southern Europe and Africa. There have been astonishingly fast rises throughout Europe in the cost of sugar, up 50%, cheese and milk, up a quarter, eggs and potatoes, around a fifth more expensive. But the UK has suffered worse because of a shortage of labour to pick produce, stemming from a decision by the government to limit immigration of unskilled people. The governments of Johnson, Truss and Sunak all interpreted – correctly – the 2016 referendum vote to leave the EU as in part a vote to control numbers of people moving to the UK for work. That is a reasonable interpretation. But it was not a vote to leave fruit and vegetables in the fields to rot and go to seed, or a vote not to plant fruits and salads in the market-garden fields and polytunnels of southern England. It was not a vote to make it even more expensive for those on lowest incomes to buy healthy food and the basics of life.

The response in some EU countries was various different forms of direct price controls. The French government negotiated with seventy-five food manufacturers to lower prices on hundreds of items. Greece put a cap on profit margins of food and essentials. Hungary and Croatia put a ceiling on the price of basics. And Spain cut VAT on food. As the *New Yorker* magazine put it, the extreme swing in the reputation of the German economist Isabella Weber is a parable of how the return of inflation is changing the politics and economics of how to manage it.* On 29 December 2021, when the world was in the grip of the Omicron Covid-19 surge, Weber

---

* 'What if we are all thinking about inflation wrong?', *New Yorker*, June 2023

– assistant professor of economics at the University of Massachusetts in Amherst – made the case for 'strategic price controls' in a column for the *Guardian* newspaper. This is what she wrote:

> *There is once more a choice between tolerating the ongoing explosion of profits that drives up prices or tailored controls on carefully selected prices. Price controls would buy time to deal with bottlenecks that will continue as long as the pandemic prevails. Strategic price controls could also contribute to the monetary stability needed to mobilize public investments towards economic resilience, climate change mitigation and carbon-neutrality. The cost of waiting for inflation to go away is high.*

This, according to the *New Yorker*, is what happened next:

> *A business-school professor called it 'the worst' take of the year. Random Bitcoin guys called her 'stupid.' The Nobel laureate Paul Krugman called her 'truly stupid.' Conservatives at* Fox News, Commentary, *and* National Review *piled on, declaring Weber's idea 'perverse,' 'fundamentally unsound,' and 'certainly wrong.' 'It was straight-out awful,' she told me. 'It's difficult to describe as anything other than that.'*

What on earth was going on? The *New Yorker*'s interpretation seems correct:

> *In a matter of hours, Weber, who was thirty-three years old, had transformed from an obscure but respected academic . . . into the most hated woman in economics – simply for proposing a 'serious conversation about strategic price controls.' The uproar*

was clearly about something much deeper than a policy suggestion. Weber was challenging an article of faith, one that had been emotionally charged during the waning years of the Cold War and rarely disputed in its aftermath. For decades, the notion of a government capping prices had evoked Nixonian cynicism or Communist incompetence. And Weber was making her case in a climate of economic fear.

Remember, all this outrage was happening *before* Russia invaded Ukraine. Inflation was already rising very significantly because of the pandemic's disruption to supply chains, how products are shipped and delivered across the world, and as a consequence of a post-lockdown surge in demand for goods, foods and energy. But it was only when the gas price rose a staggering tenfold, when access to Russian gas collapsed – initially with Putin using gas supply as a weapon against the West, and the West then boycotting Russian energy as a form of economic retaliation – that country after country chose to subsidise the cost of power to homes and businesses. Suddenly, the ideas promoted by Weber went from 'truly stupid' to totally mainstream.

The question therefore is what more governments should do to intervene in the price formation process. There is a strong case, at this time when one in twenty run out of food every fortnight* and even more risk obesity because all they can afford is sugar-rich and fat-saturated junk, to identify a limited number of healthy essentials, and subject them to the same kind of price-cap regime as applies to energy. Or at least to do that until living standards for the poorest recover.

---

* 'Impact of increased cost of living on adults across Great Britain', Office for National Statistics, May 2023

If the government should re-evaluate the use of what were once mainstream tools to limit the social hardship of inflation, there is an associated question of whether the mandate for central banks to target inflation of 2% is optimal. In the mid 1990s, I had interminable conversations with the former economic adviser to the Treasury, Ed Balls – and to a lesser extent with the Chancellor at the time, Gordon Brown – about the specificity of the inflation target, when Brown and Balls were debating whether to hand over control of the setting of interest rates to the Bank of England, prior to Labour's landslide election victory of May 1997. Brown and Balls inherited a system in which the Chancellor, not the Bank, set interest rates to maintain inflation at 2.5% 'or less'. It was an asymmetrical target. If inflation was above 2.5%, that was failure, but it was fine – under the then prevailing rules – for inflation to be below 2.5%. It meant there was an incentive to set interest rates at levels that were more likely to generate disinflation. And the logic of instituting a disinflationary bias was that above-average inflation had long been the British disease. In fact, an analysis by the Bank of England official Charlie Bean showed that inflation would be above 2.5% about half the time, if the past were a guide to the future. It also made sense to try to counteract the natural instinct of elected politicians to set interest rates at low levels that please voters.

But the whole game changes if independent central bankers take charge. They are educated – even brainwashed – to believe that nothing matters more than the soundness of the pound (or the dollar, or euro, or Swiss franc). Their instinct would always be to set interest rates a little higher than may be necessary, because their fear of inflation is so acute. Balls was therefore concerned that if the Bank of England as newly independent central bank was given an inflation target of 'x%

or less', that skew towards disinflation would mean the Bank would consistently set interest rates higher than was optimal for growth and investment. Balls was minded to change the target to a straightforward point target of 2.5%. In other words, the Bank would need to take corrective action when and whether inflation was rising above the target or falling below it.

At the time, Balls was confronted with arguments from the most senior Treasury officials – from the Permanent Secretary, Terry Burns, and the then chief economic adviser, the late Sir Alan Budd – that the removal of the disinflationary bias in the target would be seen in the City and by investors as the government lessening its commitment to fight inflation. Balls prevailed. There would be a target of 2.5% inflation, which was reduced in December 2003 to 2% – which wasn't a reduction in the target, in a real economic sense, because the 2.5% target was on the RPIX measure (Retail Price Index excluding mortgage interest payments), whereas the 2% uses today's Consumer Price Index measure, and CPI inflation tends to always be lower than RPIX. But as insurance against the Bank being inadequately rigorous in pursuit of the target, a rule was introduced that each time inflation was at least one percentage point above or below target, the Governor would have to write a mea culpa letter to the Chancellor explaining what had gone wrong, and what was being done to correct it.

Balls's system worked. Although Bean's work suggested the target would be missed and letters would be written almost half the time, there were ten years of remarkably low and stable inflation. It wasn't until April 2007 that the first such letter was sent by a Bank of England Governor, then Mervyn King.

How different from today. On 23 September 2021, the Governor, Andrew Bailey, wrote a letter to the then Chancellor, Rishi Sunak, explaining why inflation was 3.2%, or more than 1% above the target. This was five months before the energy price shock associated with Putin's invasion of Ukraine on 24 February 2022. Its then chief economist Andy Haldane had for months been concerned about incipient inflation. Thus in the June 2021 meeting of the Bank of England's Monetary Policy Committee he was a lone voice calling for the Bank's programme of government bond purchases, or QE, to be cut by £50bn – which would have represented a pre-emptive inflation-dampening rise in market interest rates.

Complacency about inflation went wider than the Bank of England's Monetary Policy Committee. In early October 2021 the prime minister, Boris Johnson, told Sky News 'people have been worrying about inflation for a very long time and, by the way, those fears have been unfounded'. This was the most perfect synopsis of what was going wrong: those in power, in Westminster and at the Bank, typically regarded anyone warning of a return of inflation as a wolf-crying boy.

Shortly after that, on 16 December, another nostra culpa letter was written by the Governor. In total there have been – at the time of writing – a staggering eight consecutive letters at the statutory interval of three months, each one of which contains a promise and a plan to reduce inflation back to target. It is what institutional failure looks like. The letter of 15 December 2022 was in response to inflation in November that year of 11.1% and that of 23 March 2023 came after consumer price inflation had been recorded at 10.4% in February.

So is it worth reopening Ed Balls's argument about

symmetrical versus asymmetrical targets? Could it make sense to change the target to '2% or less', and reinforce a disinflationary weighting? This is tricky. As I've discussed, inflation in the UK is prevailing alongside lacklustre growth and falling living standards, not the traditional inflationary boom. It's a period of 'stagflation'. Setting interest rates at levels guaranteed to wipe out inflation risks triggering a serious recession.

Is there therefore an argument that the central bank should be incentivised to tolerate a slightly higher level of inflation, say 3%, plus or minus perhaps 1.5%? Alternatively, the existing 2% target could be made asymmetric on the upside – that is inflation 1 percentage point about 2% would be tolerated, but any inflation rate below 2% would be a failure. The case for a slightly more inflationary framework is that it might help any government that wants to encourage significantly more growth-promoting investment in the UK to stimulate growth. Such an accommodating monetary and interest-rate framework might be helpful. But there is a catch-22. It is probably only practically possible to increase the target and change the framework after inflation is convincingly back at 2% or less. Because to do it when inflation is double or treble that would spook investors. They would be concerned that the target was being increased simply because the UK authorities no longer see inflation as a cancer. And in those circumstances they would dump the pound and sterling assets, which would have the effect of fuelling inflation via a worsening in the terms of trade. It would achieve precisely the opposite of what any government would want to achieve. So the only time an inflation target can be increased is after a government has tolerated the kind of monetary squeeze we are enduring now, despite the harm it is wreaking to the livelihoods of households and businesses. As Andy Haldane told me on my show this summer, there may be a case to loosen the Bank's mandate in time, but it's impossible for the government or the Governor of the Bank of England to even hint that might be sensible unless and until inflation is definitely back under control and on target. Strikingly, Sir Keir Starmer has not ruled out a Labour government altering the inflation target. There is therefore a risk, the nearer we are to a general election, that the pound weakens if Labour looks a racing certainty as the new government – which is what today's polls suggest.

In the 1970s, the inflationary shock led to no decisive political swing for years. Labour was elected in 1974, but with fewer votes than the Tories and with no overall Commons majority in a February election and then with the slimmest of majorities in a repeat battle in October. It would be five years before a succession of economic and political crises – the humiliating request by Denis Healey, the Chancellor, to the International Monetary Fund for an emergency loan to support the currency, strikes that brought the country to a halt – would usher in a decisive right-wing shift under Thatcher, and then Ronald Reagan in 1981.

In the current instability, it would be foolish to forecast the political mood five years from now. What is clear, in the UK, and perhaps in the US too, is that there has been a shift to interventionism normally associated with the left – whatever Starmer's reluctance to commit to higher taxes and higher levels of public spending. The underlying reason is disenchantment with the private sector, and in particular its perceived response to higher inflation. As I've pointed out, large companies in energy, food and banking have used the cover of rising energy and input prices, and also rising wage costs, to push up retail prices more than is appropriate. Some businesses have engaged in cynical profiteering. Others have recovered some or all of the cost increases in higher retail prices, but - arguably - could have sacrificed more of their profit margins for longer. The Bank of England has, for example, calculated that when input prices are falling, energy and food firms typically reduce what they charge us with a lag of six months, compared to the speed when they are putting up prices. It means the cost of living squeeze is being elongated for longer, and the detriment is greatest for those on lowest incomes.* Even financial analysts, such as Albert Edwards of giant French bank SocGen, have shone a light on 'greedflation' – which Adam Smith warned us to anticipate 250 years ago.

Our economic system only retains legitimacy – and its main corporate participants only retain their de facto social licence to operate – when competition leads to fair pricing and the prevention of excess profits. Profiteering is bad in and of itself, because it means living standards for the majority are lower than they should be. Profiteering is also a spur to inflation. For any government, the possible responses are short-term windfall taxes, which we've been seeing in the UK and elsewhere, direct price controls, which are also back, spurs to competition through forced break-ups of businesses with greatest pricing power, and an extension of public ownership (Starmer has backed away from the wholesale nationalisation proposed by his predecessor Jeremy Corbyn but he is planning a new state-owned green energy business, GB Energy).

The argument for the status quo is not compelling. And

---

* 'How do firms pass energy and food costs through the supply chain?', Bank Underground, August 2023

if this is year zero again in the struggle to contain and reduce inflation, it is important to note that social conditions are very different from 1974 or 1979. Then, there was significantly less income and wealth inequality. Today's inflation, savaging as it disproportionately does the living standards of the poorest, is wreaking havoc to the fabric of the nation, shredding the social contract that holds us together. This is very different from forty years ago, when inflation was the greatest source of damage to a private sector that is vital to the competitiveness of British businesses – and the long-term prosperity of everyone.

Sunak and Hunt have been antagonising doctors, teachers, nurses, railway workers and civil servants by forcing on them cuts in their living standards, by rejecting their pay demands. And they've been making themselves less than popular with most British people, whose lives have been badly disrupted by the consequential strikes in the health service, schools and on the trains. But government confrontation with organised workers over pay was the norm throughout the 1970s, even when Labour was in power. The kinds of solutions that were pursued distinctively by Thatcher, shrinking the money supply, letting interest rates soar, cutting taxes for the richest, privatising everything, would be seen by the majority – those who aren't neo-liberal zealots – as vindictive and likely to do more social harm than any economic good. We are finally and definitively in a post-Thatcher era.

★ ★ ★

PS I wanted to write about digital money, and some eccentric thoughts I've had. I haven't enough to say to fill a whole chapter, and they don't fit neatly into the arguments of other chapters. But I can't let them go. My argument is that digitisation of money may be the route to salvage the reputation and competence of central banks, in spite of the many years of cryptocurrencies over-promising and under-delivering.

Bitcoin, cryptocurrencies and digital currencies have been generating hype and disappointment for almost fifteen years. At various times there has been an enormous amount of excitement attached to them, about how they would revolutionise finance, lead to the end of conventional banks, replace national currencies, undermine the power of central banks, and somehow hand power over the financial system to people rather than faceless institutions. It was the anarchists' currency of choice. And it turns out, not enough of us are anarchists, so the crypto-induced eviction of the social and economic establishment never materialised.

There are lots of cryptocurrencies. Most of them have become fads at different times. Their exchange value, in relation to dollars, has risen and fallen and risen and fallen. Each one's story is a parable of the greed and gullibility of humans. Dogecoin started as the joke project of two software engineers Billy Markus and Jackson Palmer, and became a cryptocurrency meteor. It is a tale worthy of Maupassant. Fortunes have been made and lost, especially in Bitcoin. But they have remained on the sidelines of global finance. They've been used in the black economy, on the dark web, as a way of financing criminal activity, or to launder money, or as a means to avoid sanctions. But no one feels excluded from normal commercial life if they don't possess any crypto or digital money. To that extent this technological revolution has been a near irrelevance.

Could this change? I think it is very unlikely that transactions in cryptocurrencies will within our lifetimes finance any significant proportion of global income – except in an extreme, *Mad Max* scenario of total chaos, where all formal government and international rules collapse. You could imagine a Bitcoin or other decentralised blockchain-based currency becoming a standard means of exchange if everyone lived off the grid, if the interconnected law-based institutions and constitutions of the world were shut down. Perhaps cryptocurrencies will become the currency of choice as and when the soon-to-rise god-like Artificial General Intelligence destroys civilisation as we know it.

But within our more dull world of conventions, there may well yet emerge a commonly used digital currency, even perhaps one that doesn't respect borders in the way that sterling, dollars or yuan do. The most obvious creator of such a currency would be one of the global tech giants, whose daily operations involve billions of people across the world every day. Facebook, now called Meta, tried and failed to launch just such a currency, Libra. But it killed Libra before birth, in early 2022, because of opposition from US regulators and from senior members of Joe Biden's administration. But Meta's failure does not mean all such initiatives would be bound to fail. For example, given the vast global cash flows of a company such as Amazon, it is slightly surprising that it hasn't created its own medium of exchange, whose transaction costs would in theory be a fraction of Mastercard's or Visa's. And it would not be such a leap for Apple to transform its Apple Pay service, which currently links to a conventional credit or debit card, to some kind of account holding new digital money.

A senior central banker told me the other day that his working assumption is that there will be other attempts to create a digital currency that would become a widely accepted medium of exchange. 'My assumption,' he says, 'is that Facebook favoured the idea because it would allow them to create a form of money which enables them to harvest user information even more effectively.' Information about each of us converts into money and power for the owner. Collectively, we have all made a terrible mistake in handing so much information about our habits and preferences to the Metas and Googles for free. But until we find a way to take control of our data, and charge for it, the digital giants will continue to find ways to gather it. That is why this central banker adds that he and his colleagues assume that at some point there will be demand for digital money, that there will simply come a moment when its convenience and low cost become obvious to everyone – though he cannot forecast when that will be.

This is where it gets interesting in my view. There are three forms that a digital currency could take that would gain widespread use. One would be a so-called stable coin that largely stayed outside of the banking and central banking system. It would still have to be subject to some degree of regulation if it were going to take off. But if it was created and managed by an Apple or an Amazon, it would be as if they owned a super-efficient, very low-cost global bank that facilitated transactions and held deposits, but didn't do much else. The second version would be a digital currency widely adopted by the big commercial banks, like JP Morgan, NatWest, Barclays and so on. Essentially, they would offer digital or 'tokenised' bank deposits. This kind of digital currency could in theory become much more integrated into the entire financial ecosystem, including the economically vital function of loan or credit creation.

But if commercial banks go down that track, the question

then arises whether this new commercial bank digital currency should be a pure invention of the commercial banks, though regulated by central banks, or whether it would be better if it was effectively created by or in partnership with the central banks. This brings me to what excites me most, namely the possibility – the probability in fact – that we'll see central bank digital currencies, or CBDCs.

Most central banks have been examining the potential of these, including the Bank of England, the People's Bank of China and the European Central Bank. Like cryptocurrencies, they would exploit blockchain or other forms of distributed ledger technology, though unlike a Bitcoin or Dogecoin, the creation of new money would be centralised and subject to official permission: 'mining', or money creation, would not be open to anyone with the requisite computing power.

This kind of central bank digital currency would be 'outside' money, to use the jargon; it would be outside of the commercial banking system and of financial markets. But it could be linked to digital deposits and digital loans made by commercial banks, the so-called 'inside' money, in the way that today's bank reserves of pounds, or what's called high-powered money, relates to conventional deposits and conventional loans made by commercial banks.

Here is where all this become relevant to our current plight, namely that central banks – especially the Bank of England – were worryingly behind the curve in seeing the emergence of incipient inflationary conditions. Also, their instruments for curbing inflation have turned out to be blunt – sledgehammers rather than scalpels. But if all currency became digital, that is inside and outside money, and it was all a single currency that was part of a distributed ledger system, with every transaction leaving a digital trail or mark, in theory, the Governor of the Bank of England could sit in front of a set of giant screens in an operations room and see every transaction taking place in the economy in real time. Artificial intelligence programs could aggregate the information to make it useful and usable for policymakers, while ensuring there were no traces to personal information about any of us, to protect our privacy. They would see, minute by minute, when and where inflationary bubbles are developing, and whether these are localised or have a more general significance. Similarly, it would help the Bank to fine-tune what kind of monetary loosening or tightening would help more speedily to snuff out possible inflation or stimulate when disinflation is too strong. Even if only some pounds became digital pounds, at least some of the fog that always hangs over the economy would lift.

Such digitisation of money doesn't quite turn a central bank into air traffic control. But it would, in theory, turn monetary control and interest-rate setting into something more like science and less like hanging a piece of seaweed outside the house to forecast the weather.

CHAPTER 7
GROWING PAINS

Liz Truss, Rishi Sunak and Sir Keir Starmer have a shared obsession: to get Britain's economy growing again, after years of stagnation. Wishing it is significantly easier than doing it. In Truss's case, her impatience to restart the motor led her to almost crash the economy and get herself thrown out of the driver's seat in record time, with just fifty days as prime minister. The Treasury is still in shock about her and her Chancellor Kwasi Kwarteng's decision to announce £45bn of unfunded tax cuts, including abolition of the top rate of tax and a reversal of corporation tax rises. According to one senior Treasury source, officials acceded to this package to ward off an ambition that would have delivered an even greater shock to the economy and society. She wanted to abolish all but one income tax rate and impose what's known as a flat tax, a single tax rate for everyone. This would have been bold. It would have been the end of a progressive taxation system, one where the rate of taxation rises the more you earn. And it would have left an even larger gaping hole in the public finances in the short term, predicated on the dangerous idea that there would have been such an electric charge to the UK's growth engine that within some meaningful period economic activity would have reignited to deliver rising tax revenues. The Treasury bought it off by caving to the highly contentious abolition of the top 45% rate of tax – which so flew in the face of what even Tory MPs saw as natural justice in unequal Britain that she U-turned on it even before being forced to dump her entire tax-cutting package. In the meantime, however, the interest rate on the government's debt went from being lower than that of the US, which had been the norm for many years, to being significantly higher, as international investors decided that the risk of investing in the UK had become too much. The steady-as-she-goes risk premium enjoyed by the British government through plenty of difficult economic times was converted under Truss into what the analyst Dario Perkins of TS Lombard memorably characterised as a 'moron risk premium'. It's the albatross around Truss's neck that she'll probably never be able to shake off.

She was, of course, correct that Britain has a serious growth problem. What she, as a post-referendum zealous convert to the religion of Brexit, didn't seem to realise was that the very low rate of interest paid by the government on its debts in the immediate aftermath of the Brexit vote and for the years leading up to her tax-cutting splurge, was the market's disapproving verdict on leaving the EU. If you look at a chart of what the government pays to borrow, it fell relative to what Washington is charged as soon as the British people declared they wanted out of the EU. In economies where productivity and the underlying growth rate are higher, the 'natural' or 'neutral' rate of interest also tends to be higher, because investors are chasing a limited stock of capital and debt to invest. But the lower that any country's growth potential is, the lower the investment potential, and the lower the cost of borrowing. In that sense, Truss's maniacal mini-budget was a desperate admission of the harm done by Brexit – which the OBR puts like this:

> *The post-Brexit trading relationship between the UK and EU, as set out in the 'Trade and Cooperation Agreement' that came into effect on 1 January 2021, will reduce long-run productivity by 4 per cent relative to remaining in the EU. This largely reflects our view that the increase in non-tariff barriers on UK–EU trade acts as an additional impediment to the exploitation of comparative advantage . . . Both exports and imports will be around 15 per cent lower in the long run than if the UK had remained in the EU.**

It calculates cumulative lost investment since the EU referendum as a growth-suppressing £340bn, and forecasts the economy will be permanently smaller by 4%.

Even so, Starmer says he won't countenance resurrecting his abandoned love for EU membership, that a Labour government won't even think about trying to reverse the referendum result. So no one can accuse him of making it easier for himself to achieve his top three priorities for any government he would lead – which he declares are 'growth, growth and growth'. At the time of writing, he doesn't have a comprehensive set of policies to expand GDP, or national income, at the kind of resurrected speed that would lift all boats. He is right, of course, that it would help if the UK's planning laws weren't so restrictive – if it were easier and speedier to build homes, offices, wind farms and factories. He is also on to something important when he argues that 'greening the economy' – accelerating the transition to

---

* 'Brexit Analysis', Office for Budget Responsibility, April 2023

zero-carbon energy, manufacturing the components of wind farms and other climate-friendly sources of power, and facilitating the transition to zero-emissions of all homes and businesses – is an important part of a growth strategy in itself. The required investment would spur prosperity and reduce future expensive climate shocks – though that reduction of climate costs has to be a shared global goal, it can't be achieved by Britain alone. What Truss and Starmer both know, however, but don't say enough, is that spurring headline growth is not enough.

More important is spurring the growth of productivity, or output per hour worked, and output per worker, and thereby spurring output and income per person. There are ways, for example, of increasing the growth rate without making us on average richer as individuals, or at least not by very much. And truthfully what is the point of growing the economy if it doesn't enrich us? For example, growth can be spurred by opening borders to expand the population of low-cost workers. Arguably this was behind much of the relatively healthy growth the UK experienced after the creation of the European Union's single market, which permitted the free movement of workers across EU borders. That generated significant migration to the UK after the countries from the former Eastern Bloc became EU members. This is not to argue that the growth was pointless. It generated tax revenues that allowed the last Labour government to improve public services. It increased the profitability of businesses, which invested more and thereby improved productivity to a degree. But to the extent that cheap workers were a substitute for investment in capital equipment and software, which they were, and to the extent that a growing economic pie had to be shared out between an expanding population, then income per head grew less fast than it would have done if the same growth had been associated with a stable population. It is also relevant that this substitution of cheap labour for capital may have been a significant contributor to the UK economy's greatest weakness in the years since the crash of 2008, namely the collapse of income per head and of productivity.

This is how the Resolution Foundation put it, in their report 'Stagnation Nation'.*

> *Labour productivity grew by just 0.4 per cent a year in the UK in the 12 years following the financial crisis, half the rate of the 25 richest OECD countries (0.9 per cent). Having almost caught up with the economies of France and Germany from the 1990s to the mid-2000s, the UK's productivity gap with them has almost tripled since 2008 from 6 per cent to 16 per cent – equivalent to an extra £3,700 in lost output per person. Claims that these measures of economic progress mean little for ordinary workers are common but painfully wide of the mark. Weak productivity growth has fed directly into flatlining wages and sluggish income growth: real wages grew by an average of 33 per cent a decade from 1970 to 2007, but this fell to below zero in the 2010s. The result is that by 2018, typical household incomes were 16 per cent lower in the UK than in Germany and 9 per cent lower than in France, having been higher in 2007.*

They also show that the malaise, coupled with the inequality that is worse in the UK than in other big rich European countries, is most damaging to those with least:

---

* 'Stagnation Nation: Navigating a route to a fairer and more prosperous Britain', The Resolution Foundation, July 2022

It is ruinous for low-to-middle income Britons. Low-income households in the UK are 22 per cent poorer than their counterparts in France, meaning their living standards are £3,800 a year lower than their French equivalents.

The economic arguments around the benefits and costs of relatively unconstrained migration are complex and subtle. Since the pandemic, the shortage of unskilled workers caused by Brexit has contributed to inflation, which damages everyone, and the poorest most. By contrast, before the UK left the EU, in the era of imported low inflation and cheap money, it's arguable that immigration suppressed wages at the bottom end, for migrants and indigenous workers alike. What's uncontentious is that the UK economy became hooked on imported labour, because even though the government has 'taken back control' of its border, the rate of immigration for work has increased, with inflows of people from abroad double what they were before Brexit at just over 600,000 per year.

It's not enough for the UK to go for growth. It's got to be the kind of growth that maximises living standards. And to that extent Keir Starmer's target of making the UK the fastest growing of the G7 big rich economies is too crude. It would have been a better target to have pledged to close the productivity gap with the US, Germany and France – a gap that did close during the twenty years before the Global Financial Crisis and has since yawned open again. If Starmer wants to understand what he's up against, he'd do worse than travel back in time more than half a century, to shortly after he was born. Almost sixty years ago, on 19 July 1966, the following exchanges took place in the House of Commons between the late Tory MP, Sir Cyril Osborne, the leader of the Liberal Party, Jo Grimond, and the Labour prime minister, Harold Wilson.

Sir C. Osborne asked the Prime Minister if he will consult with the United States Government with a view to organising Anglo-American productivity teams similar to those organised by the late Sir Stafford Cripps, to ascertain why the United States industrial worker's production is two to three times greater than that of the British worker, to ascertain where the fault lies and to make recommendations to remedy this.

The Prime Minister *(Mr. Harold Wilson): Work of this kind is already in hand or in prospect, mainly through the Economic Development Committees. In these circumstances I do not think there would be any advantage in a new approach to the United States Government.*

Sir C. Osborne: *Since our economic troubles are mainly due to a small minority of employers who will not adopt the most modern methods of production, and the backwoodsmen of the trade union movement whose mule-like obstinacy will not allow them to be used . . . what steps will the Prime Minister take to solve this problem?*

The Prime Minister: *I entirely agree with the Hon. gentleman's assessment of the importance of productivity. The Anglo-American productivity teams after the war did a very good job . . .*

Mr. Grimond: *Would the Prime Minister agree that there is no great mystery about why productivity is higher in America? Incentives are greater and restrictive practices are fewer, and* *every worker has far more capital at his disposal. While the Right Hon. gentleman is drawing up the measures for tomorrow, he might bear these points in mind and bring our economy more in line with the American economy in these respects.*

The Prime Minister: *I did not think that that was a question, but I will certainly take note of what the Right Hon. gentleman said.**

This exchange confirms that successive British governments have been concerned for as long as anyone can remember about the relative inefficiency of British businesses, as measured by the average output of their employees. In other words, low productivity has been a British malaise throughout modern times, and despite pledges by every succeeding government to fix it.

The reason it matters is that improving the revenues generated by each worker is the simplest way to raise the wages of each worker. It is also possible to pay people more by reducing the share of income that goes to the owners and managers of businesses. But everything else being equal, increases in productivity are the *sine qua non* of sustainable improvements in living standards. Back in the year of England's only World Cup triumph, 1966, the British worker's typical productivity was anything but world class: it is striking that the then prime minister, Harold Wilson, did not dispute that their output was half that of their American rivals. Since then, there has been progress, but certainly not enough. According to the latest data from the UK's official forecaster, the Office for National Statistics, the productivity

---

* *Hansard*, 19 July 1966

of UK workers is about two-thirds that of Americans. Which is another way of saying we could all on average be almost 50% richer, if only we were as efficient and productive as US workers.

If you think the US has certain structural advantages, such as the huge size of its internal market and its world-beating prowess in technology, and these give them an unfair point of comparison, then it's worth considering that the productivity of British workers is 15% below that of French ones and 14% less than the average for the other six G7 rich leading nations.* This is not the kind of performance that wins World Cups.

Perhaps though more important than the international comparisons are the trends. I've looked at how much output per hour worked has changed since 1970 and the recent story is troubling. In the course of the chaotic 1970s, productivity as measured by the hourly output of workers increased 18%. That increment rose to 19% under the economically brutal stewardship of Margaret Thatcher in the 1980s, and then to 20%, as the currency chaos of the tail end of the Tory government was succeeded by the early years of New Labour. And then everything changes for the worse with the financial crisis of 2007 to 2008. In that first decade of the new century, output per hour worked increased just 11%, and then by less than 6% in the succeeding twelve years. To put it another way, the rate at which we've been enriching ourselves has fallen by 75% over twenty years. That is a staggering amount of potential income lost, and has been accompanied by disappointed expectations for a more prosperous life on a scale that leads

---

* 'International Comparisons of Productivity', Office for National Statistics, January 2023

voters to make what would hitherto have been seen as unexpected choices, like voting to leave the European Union.

The obvious point is that governments talk a compelling talk about the imperative of fixing 'Britain's productivity problem' – as it's widely known across the world – but we are resolutely stuck in the bottom half of the table of developed economies. It's worth also noting that the Liberal leader in 1966, Jo Grimond, made a point that has been true in every succeeding decade, and goes to the heart of what's wrong: 'every [American] worker has far more capital at his disposal'. In other words, the output of British workers is typically sub-par because British businesses simply don't invest enough.

I will return to why investment in the UK is relatively low. But the reference made by the Tory MP, Cyril Osborne, to the 'mule-like obstinacy' of trade unionists, who were – he claimed – blocking modern working practices, foreshadowed much of Margaret Thatcher's mission fifteen years later to break the power of the trade unions. She succeeded in the largest sector of the economy, the private sector, if not among public services, health, education, Whitehall and the rest. The UK's employment market became perhaps the least subject to trade-union collective bargaining of Europe's biggest economies. And although company bosses celebrated – and paid themselves more while depressing the share of corporate revenues that goes to employees – there was no game-changing increase in investment or output per worker. Grimond's similar complaints about 'restrictive practices' in the workplace have not been bequeathed to his successors in what's now called the Liberal Democrat party, but they are the refrain of the Brexiter Tory right, MPs such as Jacob Rees-Mogg and Liz Truss.

There is no great mystery however about why Britain's productivity has been so much lower than its competitors, and why its people are so much poorer. Among other causes, we have inferior infrastructure – roads and rail – compared to a country like France. We have fewer well-funded investment institutions prepared to invest for the long term in new businesses, through and beyond the immediate cycle, than Silicon Valley's network of venture capital funds. More generally, we have a financial system that is not efficient enough in channelling capital away from poor-performing businesses and towards those with growth prospects: too many mediocre businesses are propped up; top-class ones too often emigrate to America when they seek world domination. We have planning laws that – compared to China and America – make it hard to build new homes, new factories, and new onshore wind farms or other sources of clean green power. We have a workforce endowed with too few high-value practical and commercial skills. And we have an education system that reinforces the power of the elite while not giving enough opportunities to those from disadvantaged backgrounds.

None of the necessary reforms are difficult to comprehend. Some of them, like reforms to the education system, are easier to describe in generality rather than detail. But the big common factor linking all of them is that they are difficult to pursue to a successful conclusion in the UK's dysfunctional political system. Some, like overhauling the schools curriculum or improving infrastructure, are the work of several parliaments. So Britain's first-past-the-post political system, and the perennial risk of change in the ruling party, makes it much harder to bring about long-term change and renewal. In countries where governments that are elected using systems of proportional representation and coalitions – government by consensus – are the norm, radical changes in industrial policy rarely happen between elections. Which means the cost of capital for businesses, the cost of raising money to invest, is lower, because the environment for business is more stable and predictable.

Also, the British system where individual MPs represent a narrow geographical area, a constituency, and in which longevity as an MP depends on pleasing voters in that constituency, creates problems in pushing through policies likely to alienate those voters. An obvious example is that citizens in leafy rural areas tend to hate the idea of houses or wind farms being built in or within sight of their backyards. Predictably therefore the Tories, who have so many MPs in countryside seats, have backed away from building houses and wind farms, despite a crying need for both, whereas a more urban Labour Party says it wants much more of both.

One of the great clichés of British politics is that serious reform – of the health service, of care for the elderly and frail, of the housing market – would be possible if only the big parties would reach an agreement and stick to it, if only certain issues of basic importance could be removed from party politics. That is true, of course. And in no set of policy issues is it more true than those relating to productivity. In fact, fixing quite a lot of the other seemingly intractable issues, like housing and health, would go some distance towards improving the trajectory for growth and productivity. A healthier population, and a population whose freedom to move to new employment was not constrained by housing shortages, would be a more productive population.

This is considerably easier said than done, especially when MPs are largely split on what to do about lacklustre productivity along the lines that divided them when they were savaging each other over whether to stay in or leave the EU. Many of the Brexiters are still largely fixated on using the UK's 'new found freedoms' – as they put it – to reshape the rules and regulations that affect business so that they better capture the essential Britishness of our businesses. In many cases this has been code for stripping employees of protections, over working hours or tenure, even though there is no serious evidence that reinforcing workers' conditions and rights reduces productivity. In fact, there is an important chicken-and-egg argument here, namely, can countries like Germany or the Scandinavian countries afford their high social protection precisely because of high productivity or is their relatively high productivity a consequence of business managers' investing and reconfiguring to cope with the high levels of social protection? The important point is that responsible leaders of any decent business will – more than anything else – configure the business so that it yields satisfaction and security for the workforce. That ought to be the priority ahead of maximising returns to shareholders, despite British corporate law and custom putting the onus on rewarding shareholders. My argument would be that a contented workforce is essential to maximising shareholder returns over the long term, but there are exceptions and the two sets of interests aren't always at odds, since workers are owners of businesses through their pension funds.

There is widespread evidence that businesses that are run to maximise cash flow – rather than as a culture to provide secure and rewarding employment over the long term – don't tend to end well. Compare Sir Philip Green's defunct Arcadia and BHS groups with thriving IKEA, for example. Green borrowed to buy famous high street names, including Topshop, Burton, Miss Selfridge and BHS, sold their properties, and extracted more than £1.5bn in dividends from them. But there was never enough money invested to revitalise the more stale brands, and the balance sheet contained no financial buffer to protect businesses from the squeeze of a recessionary downturn and ongoing obligations to pensioners. What is humiliating for Green is that Topshop had a strong enough franchise when he acquired it to develop into a global giant like Uniqlo or Zara, but he wasn't patient enough to build it methodically, rather than maximising returns for himself. That's a story as old as capitalism of greed meeting a fall. But what's more important is the pernicious culture of fawning bankers and journalists who praised Green as the supposed modernising genius of fashion retailing, who got drunk at his lavish parties, puffed him up, and respectively lent him vast sums of money and the implicit backing of their newspapers. Green represents one important aspect of the British economic malaise: the belief in quick and dirty fixes, and the perennial search for business heroes who will sort it all out.

It is striking, and slightly depressing, how little of substance has changed in the financial and media culture around British business since I first started reporting on it in the mid 1980s. The City of London is – of course – much more international, and is dominated by the big Wall Street investment banks, whereas in the 1980s, before Margaret Thatcher's Big Bang, it was a closed shop – a walled garden – of British merchant banks, stockbrokers and stockjobbers. Big Bang brought vastly more money into the City, but there was no significant influx of bankers and investors prepared to be more patient when investing and waiting for returns on their money. The public relations industry around business and the City has also become more professionalised and international. When I started it was a Wild West, a semi-corrupt culture of brokers, public relations advisers and journalists getting drunk together, swapping information, helping each other out, promoting the supposed new rising stars of Thatcher's new UK of free enterprise and privatisation. There was always a story of some business – or more usually, some business genius – who would make his (almost always 'his') followers rich.

Even the government was at it. Privatisations of BT, the power generators, British Gas and so on were priced low enough to give a guaranteed windfall of up to 80% to those who subscribed to the shares being offered and who then sold immediately. It was a naked bribe to largely Tory supporters. But it was the cult of putative business geniuses who ran companies listed on the Stock Exchange, and who had the supposed ability to buy and turn around lacklustre established companies, that caught the imagination of the press, the government and – increasingly – the wider public. In retail, Ralph Halpern at Burton Group bought a series of retail chains and founded Topshop – whose inglorious end under Green I've mentioned. Terence Conran built up another high street conglomerate, Storehouse. In the industrial sector, Sir Owen Green at BTR and the Lords Hanson and White at Hanson Trust were making huge acquisitions on both sides of the Atlantic. There were many other smaller copycat versions of these ventures. Their technique was to use the currency of their shares to buy other companies, and then – supposedly – supercharge those businesses. In many cases, however, the supercharging often came from creative financial engineering, rather than significant improvements in operating performance. And at the core of these takeover machines was a very simple methodology. At the top of the business would be a charismatic entrepreneur, or in the case of Hanson Trust, a pair of likely lads, Gordon White and James Hanson – who became one of Thatcher's friends in business, and who as a young man had dated and almost married Audrey Hepburn. The alleged prowess of the Hansons, Halperns and Conrans would be sold to the City by their stockbrokers and bankers, and to the City pages of the newspapers by the cowboys of the PR world. The share prices of their respective companies would be pumped up, and then they'd use their highly rated shares – their 'paper' – to buy all the shares of less fashionable companies, whose rating was lower. As a matter of basic arithmetic, these takeovers inflated what investors regarded as a basic measure of corporate success, corporate earnings per share. It was a gravy train. Stockbrokers and merchant bankers garnered hefty fees from the deals. The executives at the top of the acquiring companies would get rich from paying themselves bigger salaries and from the huge gains on the share options they'd been granted. The managers of the target companies would rarely complain because the value of their share options would be magnified too. And City editors at newspapers had what they thought of as sensational stories about the new heroes of Britain's supposed industrial regeneration.

All that money and access to power was seductive. I recall how in my early twenties I felt as though I had stumbled into a glamorous movie when I and my colleagues on a small investment magazine, the *Investors Chronicle*, had lunch in the private dining room of London's illustrious Connaught Hotel with Hanson and White. I'd never drunk claret as delicious or expensive. And I'd never heard an executive talking so openly about his conversations with a prime minister. Hanson boasted about how he was trying to persuade Thatcher to sell him the government's substantial stake in BP – which felt to a young reporter like being in the room as the fate of one of the country's most important companies was decided, though the deal never happened. This was seemingly how Britain's economic future was decided, in private meetings of an inner circle of business people and bankers, fuelled by expensive drink and Havana cigars. One City feast, hosted by one of the smaller conglomerates that were imitating Hanson, sticks in my memory. It was a black-tie event, in a vast Victorian hall belonging to one of the City's guilds or livery companies, all gold-leaf decoration and portraits of stern old men on the walls. The main course was the predictable huge slab of beef, everyone drank themselves to oblivion on claret and Cognac, and there was a pile of Havanas in the centre of every table. From a dais at the front of the hall, the then Lord Chancellor, the late Lord Havers, gave a tub-thumping speech to inebriated adulation about what he called Thatcher's magnificent crusade to smash the trade unions. It was profoundly alienating, though instructive. Nothing about it felt modern, though companies like the host were supposedly the country's future.

Pretty much all of these conglomerates came to undignified ends, either because their alleged management prowess turned out to be an illusion, as in the case of BTR, or because they got to a size where there were no businesses left to buy that were big enough to make a difference to their earnings growth. They left a mark. Hanson's notorious squeeze on ICI, or Imperial Chemical Industries, which was regarded as the jewel of British industry, led to its dismantling. I am not sure they did serious harm to the UK as get-rich-quick schemes for their executives and backers, except that they were a substitute for a proper industrial policy that might have reshaped the economy to enhance growth prospects in a sustainable way. And this important lesson was not learned, by either bankers or politicians, because when they slipped into obsolescence, they were replaced by private equity firms – and one-man equivalents like Philip Green – who used debt to buy Britain's steady unglamorous businesses rather than the shares issued by the stock-market listed conglomerates. This was, in part, another version of financial engineering. The fall in interest rates made it cheap to borrow huge sums and buy companies whose earnings were stable. It was a bit like buying a house in the same era with a huge mortgage. If benign economic conditions, and the glut of cheap money in the world, pushed up the value of the house or the business, it could be sold rapidly for a vast profit, because even a small rise in the overall value of an asset translates into a massive capital gain if there is a big slug of debt secured against the asset (for example, if an asset is bought for £100, and rises in value to £110, there is a gain of 10% if you own the whole thing; but if you've borrowed £50 from the bank, then your £50 goes to £60, and you've made 20%; this so-called 'gearing' is a fast and simple route to make quick money in a rising market, though an equally rapid route to lose everything in a falling market).

For the past twenty years, the takeover market in the UK has been dominated by private equity buyers, such as Permira, Blackstone, KKR among many others. Sometimes they do improve the businesses they buy, when they employ very talented managers or where they see a business opportunity missed by incumbent management. And in the process they've frequently made their own executives and their investors much wealthier. But again, it is very hard to argue that they've significantly enhanced the prospects for the UK. The UK may not quite be the *Titanic* in an economic sense, but both the earlier conglomerates and the later private equity firms have been rearranging the deckchairs.

What has been missing is a systematic attempt to challenge the economic dominance of America and Germany. Which is needed more than ever now that we've left the European Union, and access to the European single market has become so much more expensive. We have neither the sources of long-term stable funding that would see significant expansion of family-controlled exporters – our version of the medium-size companies, the Mittelstand that is the engine of the German economy – nor the appetite for risk-taking of Silicon Valley in California, where venture capital firms provide huge sums to back companies pioneering new technologies. And by the way, this is not some blisteringly new insight. These structural deficiencies in the UK have been conspicuous for at least twenty-five years. There have been numerous initiatives by assorted governments to provide tax incentives for research, or funding that's matched to private-sector finance, but nothing that compares with the scale of money and know-how available in the US. Also, America's head start is so enormous – it already has so many of the world's biggest tech companies, from Google, to Meta, to Microsoft, to NVidia, to Amazon, and so on and so on – that when the founders of a younger company want to cash in, the chances are they will sell to one of these US leviathans. This is what Demis Hassabis, Shane Legg and Mustafa Suleyman did in 2014 when they sold their pioneering artificial intelligence laboratory, DeepMind, which they'd created in 2010 in the UK, to Google. It's too common a story.

The tragedy is that even when the UK succeeds in growing a world-beating tech business, like the chip designer ARM, to a sustainable size, it's usually bought by a foreign company. In the case of ARM the buyer was the huge Japanese conglomerate, Softbank. Microprocessors designed by ARM are in pretty much every smartphone on the planet. But when it was sold to Softbank in 2016, there was no attempt by the British government to block the deal, even though it was plainly a strategically important business for the UK. It would never have been nodded through in the US, if ARM had been comparably important in America. In fact, the spokesperson for the prime minister at the time, Theresa May, said: 'This is good news for British workers, it's good news for the British economy, it shows that, as the prime minister has been saying, we can make a success of leaving the European Union.' That might have been a truer statement if the word 'good' had been replaced by 'bad', and the word 'success' had been replaced by 'failure'. At the time of writing, Softbank was on course to list ARM on the US stock market at a value of up to $70bn, which would make it the biggest initial public offering in the tech industry for almost ten years. By choosing to list ARM's shares in America rather than London, Softbank embarrassed Boris Johnson and Rishi Sunak, both of whom urged the Japanese company to select the UK for registering and trading its shares.

The ARM story is just one manifestation of an ideology in the government's most powerful department, the Treasury, which since the late 1980s has held the conviction that the public sector should intervene to the minimum possible extent in the private sector and markets, that it should not attempt to pick industrial winners, and that there should be no attempt to stop any company being sold to the highest bidder. The Treasury's solution to Britain's poor management and poor productivity was to hope that when a UK business was sold to an overseas one, there would be three benign outcomes: first, that more capital would be invested in the business; second, that the investors who sold the UK business would reinvest the proceeds in other British businesses; and third, that new, usually foreign, management would take the reins and improve the said business. This approach may for a while have slowed the UK's economic decline, compared with there being no influx of new executive talent. But the underinvestment in industries that are significantly controlled by overseas interest – like water – shows its limitations. Also, the reduced desire since Brexit to invest in the UK of so many overseas multinationals with British operations, in pharmaceuticals, motor manufacturing, even in finance, shows that the nationality of ownership matters, partly because patriotism is a powerful motivator, even in the commercial world. Overseas owners of British assets have no sentimental attachment to the UK. They won't cut the country slack if its people and government make what they regard as misguided decisions – which is how many of them saw the vote to leave the European Union.

As a colonial nation, the UK was a rapacious owner of businesses and assets all over the world. But the trend in the past fifty years has been decisively towards the UK being owned rather than being an owner. In the league table of the world's biggest companies, the vast majority are American. But even France and Switzerland have three of the top fifty, compared with the UK's one. And France's three – LVMH, L'Oréal and Hermès – represent France's domination of an important global market, for high-end fashion and consumer products. The UK's sole representative – ranked forty-fourth biggest at the time of writing – is AstraZeneca, which at the end of 2021 didn't feel itself British enough to spend $360m in the UK on a new 'next-generation active pharmaceutical ingredient manufacturing facility for small molecules' and instead decided to put the new plant in the Republic of Ireland – because, according to AstraZeneca's French chief executive Pascal Soriot, the tax rate in the UK is too high. The next biggest 'British' company is Linde, the world's largest industrial gases company, which was founded in Germany, has its operational home in Woking in the UK and its tax home in Dublin. So even those multinationals that are technically British seem rather more to be citizens of the world than their American and French counterparts.

Which perhaps would not matter if there was a thriving system of incubating and growing powerful new businesses. And compared to much of Europe, though not to the US or China, the UK has been a breeding ground for tech. Johnson and Sunak, for example, have both routinely boasted about how the UK has more 'Unicorns' – or privately held tech businesses worth more than a billion dollars – than other European countries. Sunak attracted gentle ridicule however when he said to business leaders in April at a conference called Business Connect that his ministers had been in Silicon Valley rebranding the United Kingdom as 'Unicorn Kingdom' – which sounded too much like the UK as a place where government lives in a fantasy world. One influential tech investor, Ophelia Brown, the founder of Blossom Capital, warns of the risks of not seeing the competition in a clear-eyed way:

> *For more than a decade London has been the jewel in the crown of Europe's growing tech sector. It is now almost two years, though, since my venture fund, Blossom Capital, saw a single tech company here that we have wanted to invest in. Momentum*

is ebbing away from the UK capital faster than from other cities ... The total number of investments made in London in 2021 was 2,400; a year later this figure was 1,800. The number of pre-seed investments made in 2022 collapsed by 43 per cent, while seed investments fell by almost a quarter in 2022 from the previous year. Over the same period the value of VC investment in France climbed by 10 per cent to £12.7 billion, its strongest year yet.*

So what does she think has gone wrong?

For the past decade London was the place where global tech investors came to find the next most promising companies. Index Ventures, Accel, Lightspeed, General Catalyst, A16Z and Sequoia have all built funds here. Now US investors associate the UK with economic and political crisis. These funds may be based in London but the dollars they invest here pale in comparison with those they deploy elsewhere ... If London becomes merely a departure lounge for Paris, Berlin, Warsaw, Oslo, Copenhagen and Amsterdam, that will not create jobs, nor will it help to build wealth, level up the regions or increase the tax take ... The government's ambitions to create the next Silicon Valley in the UK look fanciful. We have become a country that forces talent away at the border. The UK's biggest and best known unicorns – Revolut, Deliveroo, Checkout, Wise – were started by people who moved to London from elsewhere so this really matters. Right now, to serious-minded investors, the UK's plan to become a tech superpower looks risible.

---

* Ophelia Brown, 'London needs to up its game if it is to stay tech capital', *The Times*, 2 May 2023

Those ambitions shouldn't look laughable given the natural advantages enjoyed by the UK, in language, time zone and best-in-class universities. But another deficiency, or perhaps an opportunity if it can be corrected, is in the shortfall compared to other countries in how much we spend on research. An important review of the UK's research capability was carried out at the request of the Business Department by the Nobel prize-winning scientist Sir Paul Nurse, who runs the world-renowned genomics research centre, the Crick Institute. Published in March 2023, the review found that the government funds and carries out less research than many competitor nations, and that the commonly held view that the gap is largely in the private sector is incorrect.* He said that in research funded by the government, the UK ranked twenty-seventh out of thirty-six rich OECD nations:

> *The governments of research intensive and economically successful nations such as the USA, Germany and South Korea invest 0.66% to 0.96% of GDP compared with the UK's 0.46%. Therefore more UK Government investment in RDI [research, development and innovation] is required, investment that needs to be embedded in a stable government policy environment.*

Business too invests less in research than in some important competitor nations, though in this case the gap turns out to be smaller than was widely thought, following a review of the numbers by the Office for National Statistics. According to the ONS, business spending on R&D is equivalent to 2.4% of national income or GDP, compared with 3.14% in

---

* Sir Paul Nurse, 'Independent review of the research, development and innovation (RDI) organisational landscape', March 2023

Germany, 3.45% in the US and 2.35% in France.* This changing picture of where the R&D flaws lie has left the government struggling to know quite what to do. In the autumn of 2022, the Chancellor cut R&D tax credits for most small tech companies, before partly reversing those cuts for those spending more than 40% of their budgets on R&D. He also allowed full 100% 'expensing' against tax for investment in plant and machinery, for three years, to help soften the effect of an increase in the corporation tax rate from 19% to 25%.

This time limit on providing the increased tax break for investment was forced on the Chancellor by his fiscal rules, in the sense that it is expensive and if he had promised to roll it forward indefinitely, he would have broken his commitment to reduce public-sector debt as a share of national income in five years. But it makes the allowance less than ideal, in the sense that much investment in new plant and machinery is a commitment over years, and companies are much less likely to embark on such a long-term project if tax subsidies are set to be withdrawn and the net cost would therefore rise. This point is made compellingly by the welfare economics think tank, the Resolution Foundation, which argues – in a report co-funded by the Nuffield Foundation – for a fundamental overhaul of corporate taxation, to support investment.

The UK has discouraged investment through both policy design and instability, most recently via the welcome new policy of letting firms pay for plant and machinery out of pre-tax profits (i.e. full

---

* 'Gross domestic expenditure on research and development, UK: 2020', Office for National Statistics, November 2022

expensing) being announced as only temporary (it is set to end in March 2026). The Government should immediately make this permanent: the objective is to increase the level of investment and not just bring forward its timing. Ideally, all investment, including buildings would be paid for out of pre-tax profits.*

The Resolution Foundation says such a significant increase in the effective subsidy for investment would be expensive. It says the costs in lost tax revenues could be offset with another reform, which would be to introduce strict limits on how much interest any firm could deduct from taxable profits. I've argued for such a reform, to reduce the tax advantages of financing corporate transactions with debt, for fifteen years. It would go some way to reducing the incentives for financiers, like Philip Green or private equity firms, to load up the businesses they buy with debts, and thereby put those businesses more at risk. It would also increase the relative advantages of equity funding, and would encourage investors to take a more patient, long-term approach to seeking returns.

Another source of the UK's economic underperformance is that investment by the government and public sector is significantly less than in other rich OECD countries. Had UK public investment been at the OECD average as a share of GDP for the past couple of decades, there would have been £500bn of additional cumulative money invested. Even under Sunak and Hunt, who understand the economic benefits of investment, there has been a significant cut: in November 2022, Hunt froze public sector net investment in cash terms till 2027-8. That is equivalent to a real or

---

* Broome, Corlett and Thwaites, 'Tax planning: How to match higher taxes with better taxes', Resolution Foundation, June 2023

inflation-adjusted cut of £15bn and would see investment fall from 2.5% of GDP to 2.2% - which would be 40% below the OECD average. This damages our quality of life by weakening the fabric of public services and reducing our prosperity. According to the Resolution Foundation weaker investment - both public and private - contributed 'around 9 percentage points of the roughly 20 per cent fall in GDP per capita from its long-run trend since the financial crisis.'[*]

Perhaps the defining feature of the UK's productivity problem is that it does not apply to all companies. Analogous to the grotesque income and wealth inequalities in the UK, there are extreme differences in the productivity of British companies. This cancer of multiple forms of inequality is a theme I explored in depth in my previous book, *WTF?*. The economist who identified this pernicious productivity differential between a largely southern premier league and the rest was Andy Haldane, formerly of the Bank of England. In an article for the Royal Society, he wrote:

> *Where the UK diverges markedly from other mature economies is in the presence of a large underperforming group of less successful firms – the so-called 'long tail'. While productivity in elite companies continues to march ahead, national averages are weighed down by this long tail of mediocrity.*[†]

The country boasts some businesses that are among the most productive in the world. This is true of the financial sector, pharmaceuticals and many creative businesses. This

---

[*] Odamtten and Smith, 'Cutting the cuts', Resolution Foundation, March 2023

[†] Andy Haldane, 'Better education, better productivity', The Royal Society, June 2022

productivity gap between the best and worst is part of the explanation for the income and wealth gaps between North and South: so many of the high productivity businesses, in investment banking, life sciences, media among others – the employers that can afford to pay high salaries and other forms of lavish remuneration – are in London and the South. Too many of the glut of businesses with substandard productivity are located in the poorest parts of Britain. There is a related weakness, which is that UK managers and management practices are typically not up to best international standards. An important international study of hundreds of medium-size firms from across the world, the World Management Survey, shows that British companies typically earn significantly less from the same capital and kit, such as information technology, than their US counterparts. These lower returns 'could be almost entirely explained by the different forms of managerial capability.'* Or to put it another way, British bosses aren't good enough.

Much of the solution is implicit in the cause. The success of the City is in part because so many top-quality businesses are packed closely together – and have been for hundreds of years – so that those who work in them can and do learn from each other, almost without noticing. These are the rewards of being part of an industrial 'cluster'. There are similar networking benefits in London-based advertising and media, technology around Cambridge, life sciences in London, Cambridge and Oxford, and so on. This causes a dilemma for any government. On the one hand, there would obviously be a massive productivity gain for the UK as a

---

* Van Reenan et al, 'Improving productivity through better management practices', *LSE*, May 2002

whole if the removal of planning restraints, for commercial and residential property, were to allow the creation of a substantial tech and life sciences corridor linking Oxford and Cambridge. Quite apart from 'nimby' objections from residents, the other concern is that this would simply reinforce inequalities with the Midlands and North. The imperative therefore is to equip metro mayors and devolved governments with the financial and planning muscle to stimulate the development of other powerhouse clusters all over the UK, around Manchester, Newcastle, Belfast, Newcastle, and especially in the poorer regions.

Another quintessentially British paradox is that the country is perhaps the best in the world at coming up with huge intellectual breakthroughs, of enormous commercial potential, and then routinely and systematically failing to exploit that potential. We are best in class at the 'R', for research, in R&D, having won more Nobel prizes per capita than any other nation (as I've already mentioned). The problem is with the 'D' for development. The UK leads on innovation, but lags on exploiting or diffusing the innovation. A British genius, Tim Berners-Lee, invented the internet, but no mega digital firm is British. Another British genius, Alan Turing, was a pioneering prophet of cryptography, computing and artificial intelligence. Pretty much all his pioneering ideas have been commercialised by American companies. Most of the scientific breakthroughs in understanding the human genome were made by British scientists, including Paul Nurse, and starting with the discovery of the double helix by Crick and Watson. However, the three biggest gene sequencing businesses in the world are American. I could go on.

Successive governments have tried to solve this by encouraging universities to become more commercially minded, and – following the example of US ones, like Stanford and MIT – create a culture of academics converting their ideas into businesses. There have also been various different attempts to create assorted state-funded business banks. The British Business Bank is the latest. It invests in venture capital funds that then back potentially high growth businesses in digital technology and life sciences. For example, it stands behind British Patient Capital, which was launched in 2018 with £2.5bn of government funding to invest over ten years, following a Treasury review that identified what was glaringly obvious, that some companies were being held back by inadequate access to long-term risk finance. The British Business Bank wants more money and more freedom to reinvest any gains it generates on its investments. This is sensible, but not ambitious enough. British Patient Capital's resources are a tiny fraction of the assets under management of a single Silicon Valley VC investor, like Andreessen Horowitz or Sequoia. So, whether or not British Patient Capital is the best vehicle, the government has to take far more risk in providing equity funding and long-term finance to the kind of young businesses that have exciting proprietary technology.

When it comes to productivity differences, access to and the nature of finance is a big thing. Infrastructure in the form of cheap power, the rule of law and decent transport links also matter – which is observable in France and China. But the nature of education is pretty much everything else. This takes us back to the risks and rewards of the coming artificial intelligence revolution. It is obvious that every young person in every school should right now be taught how to make the most of generative AI models, whether they are engaged in traditional word-based research, designing graphics or writing code. The faster this becomes second nature to them, the more valuable they will be to future employers. More than that, the curriculum has to change fundamentally, to focus far less on learning facts, and far more on acquiring the critical skills to be able to challenge an AI result when it's wrong or misleading. And as I said at some length in *WTF?*, increased time and resources need to be devoted to creative disciplines – music, poetry, literary writing, the visual arts – to the pursuits that are too human for AI to mimic in anything but a soulless way. Also, and as a stark contrast, all schools should be obliged to provide proper financial education. Not just how to set up and manage a bank account, but how to create a small business, how to be self-employed and how to invest. It is scandalous that the ability and confidence to invest directly in the stock market or in unquoted businesses, including in start-ups, is typically restricted to the offspring of the rich. That famous film of a pre-pubescent Jacob Rees-Mogg saying how much he enjoys looking at the prices page of the *Financial Times* should not be an eccentric anomaly in any society that actually cares about 'levelling up'. It shouldn't only be plutocrats who spawn new plutocrats. In a radical version of the democratisation of financial know-how, a government would revive the Child Trust Funds that were originally created by Tony Blair and Gordon Brown, and killed by David Cameron and George Osborne. Each fund would be seeded by the government, perhaps to the tune of £1,000. And the money would automatically be 50% invested in a state-owned venture capital fund, and half in the stock market. When the relevant child turned eighteen, they would take control of the management of the fund and would have discretion to invest it as they saw fit, for a further ten years, including of course a safer default option. In theory, this would help all young citizens understand the principles of investing and wealth creation, and would endow everyone with a useful nest egg aged twenty-eight.

Of course, given the pace of change, young people should be helped early in a life lesson that was never discussed at school or – as far as I can remember – much at home, when I was young, which is how to fail and pick yourself up. Like many of my generation, my mother's way of encouraging my sister and me to do our homework was to shout at us, 'Do you want to end up working in Woolworths?' (For those of you below a certain age, Woolworths was a precursor to the various anything-for-a-pound shops.) Her view was that you were either someone like my dad, who was the first in his family to go to university and then had secure employment as an academic his entire life, or you were doomed to work with your hands. The kind of significant changes of employment that both my sister and I went through was more than even she would have expected. But those at school today may typically live to 100 plus, according to human biologists, and may well be working into their eighties. In their active lives, they are likely to have to change their entire careers, possibly several times. Which means governments have to turn that most empty of political pledges, 'lifelong learning', into educational services that can convert an accountant into a stonemason (which I mention, because fixing the gargoyles of an English Gothic cathedral is unlikely to be handed over to a robot ever).

As for today's young people, they must unlearn the self-harming prejudice of my generation, that compelled career change represents failure. Instead, change should be seen as an exciting challenge, as a route to a richer life, emotionally and intellectually. The nature of the welfare state will therefore have to change, so that when anyone is forced to step back from work for the weeks and months of retraining, they'll receive a decent income for pretty much as long as it takes. In *WTF?* I promoted the case for a universal basic income, which would be everyone's right, if the obstacles are insuperable – as they may be – to ending the winner-takes-all nature of economic life in a digital and AI age. That said, it may be impossible to design such a state-funded floor to permanent incomes at an affordable cost, without destroying important incentives to work. But a universal retraining grant, set at a high enough level to support a family, would seem essential to avoid the worst of the social dislocation from the emergent AI society.

CHAPTER 8
BYE-BYE BOOMER

Since at least the housebuilding boom of the 1930s, the British obsession with owning a home has dominated the national conversation and constrained the ability of prime ministers and chancellors to engineer huge economic and even constitutional shifts. A quarter of a century ago, when there was still an active question whether the UK would give up the pound for the euro, the decisive factor against joining the eurozone in the mind of Treasury officials and of Gordon Brown was the structure of the British housing market. With so many people funding the ownership of their houses with variable rate mortgages, and with interest rates in the eurozone half those of the UK, adopting the euro in the UK, and importing those lower interest rates, would inevitably lead to an unsustainable house price boom. And since booms are followed by devastating busts, Gordon Brown and his main economic adviser, Ed Balls, were set against joining the single currency long before the completion of their official evaluation, their notorious five economic tests. In their minds, Britain's housing market, and the idea that owning a home was the essence of being British, meant the country flunked their first test – 'Are business cycles and economic structures compatible so that we and others could live comfortably with euro interest rates on a permanent basis?' – even before they asked it.

Today's housing market crisis is decidedly made in Britain. Mortgage rates through the roof, prices falling, and yet younger people still struggling to get on the ladder. And although owner occupation has been falling since Brown and Balls opted to put the perceived interests of homeowners above the project of European integration, the rental market has provided an inadequate alternative: too often private and social landlords get away with charging for slum conditions. What was our national pride for so long, the democratisation of property ownership, should now be a source of national shame.

Within living memory, the housing market has had an impact on our lives more than any other segment of the economy. As a stereotypical Boomer, I was able to buy my first two-bedroom home – a flat above a sandwich shop five minutes from Borough underground station just south of the river – in the mid 1980s, for two times my modest salary. In the succeeding decades, I traded up twice, when I had a family, and then sideways, when children left home. The decision to own gave me common interests and common cause with millions of compatriots, however different our other attitudes and convictions. When house prices rose, we were happy; when interest rates rose, we grumbled. Mostly we were happy, as prices rose, and rose and rose. When we flipped the coin, we won. The losers are Kish and his generation, the Millennials. Most haven't had the salaries or wealth to get on the ladder. And when like Kish they've recently managed to do so, they wonder whether they've made a terrible mistake, as mortgage rates are reset at cripplingly high levels and selling prices fall below what they paid. An

Englishman's home has become his economic prison. So is there any chance the bust, which Brown and Balls were so desperate to avoid, could lead to the kind of fundamental reform that might reintroduce fairness and equity? There is a precious opportunity, but it requires the kind of rebalancing within the economy that I've described in earlier chapters. In particular, growth and productivity have to be revived, such that capital gains on housing are not seen as vital to offset stagnating wages and incomes.

In our recent era of flat living standards, one of the few reliable routes to increased wealth after the financial crash was from owning property, as falling interest rates fuelled the asset-price surge. The average UK house price in January 2008 was £185,000. In January 2023 it had risen to £288,000, an increase of 56%. In the same time frame, disposable incomes for those in the private sector have risen only fractionally and have fallen considerably in the public sector. For decades, owning a home wasn't just about keeping the rain out. It was about the only sure-fire way to accumulate a nest egg. You were nuts if you didn't scrimp and save to buy a place of your own. Except that, with house prices soaring and wages stagnating, increasingly they have become unaffordable for almost anyone below middle age. As the investment management firm Schroders points out, the average house in the UK currently costs around nine times average earnings, a historic high, and 'the last time house prices were this expensive relative to average earnings was in the year 1876, nearly 150 years ago.'*

If you live in Britain, failure to get on the property ladder

---

* 'What 175 years of data tell us about house price affordability in the UK', Schroders, February 2023

has left you and your family poorer and – even after the recent interest-rate rises – more vulnerable to economic shocks than property owners. This is partly why critiques of Britain being too obsessed with home ownership are correct though naive. There is a rational case that we'd be in a better place as a country if housing were less important to individual wealth, and if instead we focussed on making sure anyone could be confident of a decent home whether they own or rent. But the argument that reforming the rental market would be enough reinforces the current disparities of wealth and power. We do need a less exploitative rental market. But we also need the ladder of affordable ownership to be accessible by many more.

Even being able to think about housing needs in terms of financial returns is proof of privilege. In a series of investigations, *ITV News*'s Dan Hewitt has shone a light on the stories of those who pay the highest price for the lack of affordable housing. Whether they are living in criminally substandard social accommodation or having to deal with the worst of the private rental sector, their experiences shame a rich country. Poisonous mould, damp, leaks of water and sewage, collapsed ceilings and vermin infestations, these are the conditions that appalled the nation when George Orwell exposed them in *The Road to Wigan Pier* in 1937. They are with us still. The complacency of landlords, in the social and private sectors, is scandalous. Thanks to the investigations of Hewitt and the young housing activist Kwajo Tweneboa, who has built a powerful online platform exposing substandard housing, we can no longer plead ignorance to the disgraceful living conditions from which vulnerable people struggle to escape. In *WTF?*, written in the aftermath of the Grenfell Tower fire in 2017, I wrote that we must ensure 'Grenfell changes everything.' My and others' words have been heeded only in the most superficial of ways. Addressing the scandal of unfit homes for those who have least is the first priority in housing reform.

There are modest grounds for hope. On a visit to the Rochdale Estate in which two-year-old Awaab Ishak died from exposure to poisonous mould, the Housing Secretary, Michael Gove, accepted that his government had for too long failed to protect those who need the safety net most. He has, finally, passed the Social Housing Act, which will upgrade regulation, lift housing safety standards and give more power to tenants. It's been promised since Grenfell. Better late than never. Even though 61% of British people cite housing as one of the most important issues facing the UK, according to the ONS in August 2023, there have been six different housing secretaries since the fire six years ago. The current incumbent is in his second stint and managed belatedly to enact changes that had been promised in a White Paper since 2020.

Gove also concedes that there needs to be more decent social housing, which is more a statement of the obvious than of bold political choice. Just over 7,500 affordable homes for social rent were built in 2021–22, against 1.2 million households waiting for social housing. And because of sales, there has been a net decrease of available social homes.* After New Labour came into government in 1997, and even with Thatcher's implementation of 'right to buy', which saw large numbers buying their council homes, housing rented from local authorities or housing associations was comfortably

---

* 'Live tables on affordable housing supply and social housing sales', Department for Levelling Up, Housing and Communities, June 2023

the second most common type of residence in England, at 22% of properties. Private rental was 10%. And a massive 67% were owner-occupied. By 2010, when they left power, this hierarchy was shifting: owned property was 65%, social housing was down four percentage points to 18% and private rental had grown to 16%. Today, social rental has fallen slightly below private, though both are around 17% of the total, and owner-occupied is 64%.* Labour's Lisa Nandy, until recently their Shadow Housing Secretary, has long insisted turning this around would be a priority for a Labour government. She described their priorities for housing policy at Labour's annual conference with a Blairite mantra of 'council housing, council housing, council housing'. She wants to increase the numbers in affordable social housing back above the private rented sector, to reverse a trend that was in full swing under Blair and Brown, even if it had its origin with Thatcher.

Even the future king, Prince William, has spoken of an ambition to provide social housing on his 130,000-acre Duchy estate. He recognises that it should be our common, collective cause to increase the stock of affordable housing for those on lower incomes. One implication is that we've reached the

---

* 'Dwelling stock by tenure, UK', Office for National Statistics, November 2022

1979: 11,520 own (55%), 2,385 priv (11%), 6,568 LA + 353 HA (33%) = 20,826 total

1990: 15,099 own (66%), 2,123 priv (9%), 5,015 LA + 702 HA (25%) = 22,939 total

1997: 16,215 own (67%), 2,361 priv (10%), 4,273 LA + 1,132 HA (22%), 132 N/A = 24,113 total

2010: 17,463 own (65%), 4,385 priv (16%), 2,220 LA + 2,562 HA (18%), 66 N/A = 26,696 total

2022: 18,347 own (64%), 5,434 priv (17%), 1,987 LA + 2,939 HA (17%), 33 N/A = 28,740 total

limits of 'right to buy'. If social or council housing is a ladder to ownership of a property in the private sector, that is splendid. If it's a ladder to ownership of a high-quality property explicitly built for a social purpose, that is no longer appropriate. Building a consensus around fixing social housing – improving availability, affordability and quality – is doable. A motivated government could achieve a great deal. Fixing the private market is harder, because the conflicting interests are so powerful. The nutshell remains a spectacular mismatch between supply and demand.

On the supply side, Britain has not built enough houses for years. Estimates by the National Housing Federation and Crisis suggest that we need around 340,000 new homes in England alone each year, of which 145,000 have to be 'affordable'. That is in itself higher than the Tories' manifesto target of 300,000. But actual provision always falls well short. Last year there were 232,000,* a relatively good performance. In 2012–13, just 125,000 were built. And although there has been progress, the most delivered in a single year since has been 243,000. Until there is closer alignment of provision to need, people will be forced to put up with substandard accommodation and prices will remain too high.

Ask most experts what has gone wrong and you get one answer: stupid politics. It is not just nimbyism – no building in my backyard if you want my vote. The planning system gives veto force to the outrage of older owners, usually Tory supporters, at any development that might impair the value of their most important asset. Planning is a wholly non-strategic, case-by-case discretionary process, which creates

---

* 'Housing supply: net additional dwellings, England: 2021 to 2022', Department for Levelling Up, Housing and Communities, November 2022

uncertainty and means even socially valuable projects end up being blocked. There is always a negotiation between local authorities with planning responsibilities and developers. One widely urged reform would be for 'zoning', to specify which kinds of development are appropriate to different locations. Under this approach, if a development met the criteria for a relevant zone, planning permission would almost automatically be granted. It worked in New Zealand – admittedly a country with more wide-open spaces – and is recommended by the Centre for Cities think tank. More generally, nothing is going to sort the housing crisis without bold reform of the planning system, and this government has proved incapable of doing it. Much is promised in general terms by Labour, though at the time of writing it has provided little detail of what specifically it would do.

It is a particular disgrace that too many expensive properties are built and then sold to foreign investors, who leave them empty. So the UK's case-by-case discretionary planning is especially cancerous when it comes to affordable housing. Among the 232,000 houses built last year, just 59,000 were classed as affordable,* only 41% of perceived need. Developers and landowners are too successful at arguing that their project would be unviable if they include a substantial volume of low-cost housing. This is a clear manifestation of the kind of market failure where developers will need to face the threat that their land will be expropriated by the public sector if they don't develop it and don't do it in a socially useful way.

The dysfunction in the housing market is just as damaging

---

* 'Social rented housing (England): past trends and prospects', House of Commons Library, August 2022

to vulnerable people's welfare as the dysfunction in the energy and water markets. We take it for granted that businesses operating in energy and water are subject to stringent service and investment targets, and centralised regulation of their allowable profit margins and returns on capital. There is no good economic reason why there shouldn't be an equivalent national regulator for the major housebuilders, given their strategic importance to our prosperity and welfare, and that they've been failing to build what the country needs. They may not be the kind of natural monopolies that are typically subject to that degree of consumer-protecting intensive regulation. But once they acquire a significant area of land, or rights to develop such land, their decision whether to develop or passively hold is make or break for a community and the surrounding region. We also need to see central and local government much more active in identifying areas where significant developments are needed, and then ruthlessly forcing them through.

This is not to say that we should concrete over the country's green and pleasant land, that protecting most of the Green Belt isn't sensible, or that developments shouldn't be on brownfield sites, those already subject to earlier construction, wherever possible. But government needs to think and act more strategically in addressing housing need. And it needs to encourage metro mayors and local authorities to create ambitious building plans for their regions and communities. This should involve the transfer of financial resources to make a reality of those plans, including the resources for public ownership. This is as much about economic regeneration as it is about social policy, because the prospects for vibrant industrial clusters to expand are enhanced when skilled people can live and work locally.

In contrast to the painfully hobbled supply side, demand for houses has been surging for years. The causes are several: there is less natural turnover as people live longer, the size of household units has shrunk, and – in case you missed it – there has been a bit of immigration. Demand has also been stoked by assorted government schemes to help first-time buyers. The amount of money chasing a relatively fixed stock of properties has been increased by stamp duty holidays, credit to help with deposits, and subsidised savings. Rising demand at a time of cheap credit equals asset inflation. And then it turns into a pyramid scheme, as those who missed out on the early price increases pile in at the second and third and fourth rounds.

★ ★ ★

Our housing dysfunction does not just harm the young financially, but also distorts their plans and expectations for life in the widest sense. The most recent census data available shows five million adults live at home with their parents, an increase of 15% in the last decade. According to the ONS, in 1997 the most common arrangement for an 18- to-34-year-old was to be living with a partner and one or more children. Twenty years later, people of the same age were by a clear distance more likely to be living with their parents. This is largely about the unaffordability of housing. In areas where house prices are closer to average incomes, staying at home is less common. In fact, in the cheapest region to buy or rent in the UK, the North East, there was actually a small decrease in the proportion of adult children living at home: numbers of new family households grew faster than numbers of households with adult children. The least affordable region of the UK in which to buy housing, London, saw the fastest increase in numbers of adults boarding with Mum and Dad.

One corollary is that the average age of a first-time buyer is thirty-four years old, older than the average age of mothers having babies, which is thirty-one, according to the ONS. A quarter of a century ago, first-time buyers were twenty-six on average and mums had the first child at twenty-eight. This represents quite a radical change in the chronology of growing up. For politicians and economists who are anxious about Britain's falling birth rate, the housing question should be front and centre. The Conservative MP Miriam Cates, for example, points out that 'in 1941, fewer than 20% of women reached their thirtieth birthday without having given birth to at least one child. Now it's over 50%.' If this is largely due to changes in what women want to do with their lives, it would be benign. But Cates worries about who is going to pay taxes to support her and the rest of us in old age, if births continue to fall below the so-called 'replacement rate' of 2.1 children per family – though she could of course encourage her party to be less hostile to immigration. Presumably she is sympathetic to young people who do not want to raise a family from their childhood bedrooms, for all the joys of being surrounded by the love and support of extended family, and the tantalising prospect of free childcare.

Those who can't afford ownership will often opt for the private rental sector. But in doing so they may deny themselves the opportunity of ever being able to save enough to pay the deposit on a purchase, and they would expose themselves to instability. Families in short-term rented properties often find their rents being increased to crippling levels that price them out of their areas. This is a nightmare for any parent with children in nurseries, schools or taking advantage of local support networks.

Britain's housing problem is not just about how hard it is for the young to buy. Since the financial crisis, the percentage of income spent on rent in England has been above one-third. For owner-occupiers over this period, the proportion spent on a mortgage has been significantly lower, in the high teens – though at the end of 2022, with mortgage rates rising, it had risen to 22%.* Adding insult to injury, average floor space per tenant has declined by almost a fifth over the last twenty years, whereas those who own have seen their living space increase.† Poor housing standards and rapacious landlords are the daily experiences of younger people. Some of Kish's friends have, after ten years renting in London, confessed to questioning their faith in capitalism. As they see more and more of their income swallowed by rent for insecure and shoddy accommodation, they think Adam Smith's invisible hand is giving them the bird.

The dramatic recent increase in mortgage rates also means those who bought a home only recently are starting to experience the anxiety that those without any assets have had for some time. Data from the ONS for August 2023 showed that 47% of those paying rent or a mortgage reported their rent or mortgage payments had gone up over the previous six months, and 38% were finding it difficult to keep up the payments, up from 29% a year earlier. As mortgage rates are reset at three or four times what they were only eighteen months ago, and as house prices fall, newer owners face the

---

* 'English Housing Survey 2021 to 2022: headline report', Department for Levelling Up, Housing and Communities, December 2022
† 'Housing Outlook Q3 2022', Resolution Foundation, September 2022

potential double indignity of neither being able to afford the costs nor having any equity or wealth in the house if forced to sell. These mortgages come up for repricing at the rate of about a million a year. This means that so long as interest rates are on a rising trend, or if they stabilise at a high level, there is a progressive withdrawal of spending from the wider economy, and it becomes harder and harder to restart growth. In a possible worst case, large numbers of people abandon any hope of servicing their debts, losses escalate for the lenders, and there's a credit crunch. We're not there yet; the Bank of England is hopeful we won't see a housing crisis become a banking crisis. But it's not impossible. The economist and former civil service official Simon French noted in June 2023 that the average interest rate on outstanding UK mortgages was 2.7%, only up slightly from the 2% low of 2021. The full effect of rising interest rates will only be felt when the full refinancing cycle has played out over the coming three years.

What is to come will be especially painful for those heroic young people who have somehow managed to buy their own homes in recent years. Older homeowners have typically paid off all or part of their mortgages: they're likely to be in the 8.2 million people who own their homes outright, rather than the 7 million paying mortgages. They are more likely to be considering downsizing than upsizing in the future and are less likely to be squeezed by significant childcare and other costs associated with dependents. Older cohorts are also the beneficiaries of many years of house price appreciation. Today's price dip will reduce their windfall gains, not wipe them out. By contrast, young owners face the looming pain of negative equity, or the value of their properties falling below the outstanding mortgage debt secured on it. When it comes to monthly payments, tighter money and higher interest rates have come at precisely the worst moment. They received no interest on their meagre savings as they sought to build up the capital to pay the deposit on a house. Now, finally, that they have huge mortgage debts, the switch in interest rates is a curse. Even with more stringent stress-testing by lending banks, imposed by the Bank of England after the 2008 crash, the housing analyst Neal Hudson has calculated that the average loan taken out to buy a house is equivalent to just under three and a half times people's incomes, up from double in the 1980s. He calculates that due to the sums of money now being borrowed, the 6% interest rates on many mortgages today is inflicting the equivalent pain of 13% rates forty years ago.

With prices and rates high, those needing to reduce monthly payments are borrowing for longer and longer. Mortgages, traditionally taken out for twenty-five years, have been extended to more than thirty years for over half of new first-time buyers, and in March 2023, 9% took out thirty-five-year mortgages. Less than twenty years ago, fewer than 2% borrowed for thirty-five years.* With the average age of a first-time buyer at thirty-four, these younger buyers will still be paying the bank for years after they're sixty-four. Little chance of grandchildren Vera, Chuck and Dave on their knee. They'll still be working.

As I said earlier, there is a strong argument for welcoming an era of higher interest rates and lower house prices. This would give young people both better returns when saving for a pension and a better prospect of being able to afford to buy a home. So surely we should say, 'bring on the crash'. The problem is the one Brown and Balls identified when deciding

---

* 'Household Finance Review: Q1 2023', UK Finance, June 2023

the UK could not join the euro. A house price 'correction' is always economically benign in the long term, and a political and economic disaster in the near term. If our leaders thought they had the backbone to tough out a residential property slump, they've reassessed after the Tory reputation for competent economic governance was shredded by Liz Truss's and Kwasi Kwarteng's mini budget debacle in the autumn of 2022. No prime minister, least of all a Conservative one, wants to be blamed for wiping out the wealth of millions of older voters, whatever the long-run social benefit for the young.

As for Rishi Sunak, he has history propping up the housing market. While Covid-19 infections were raging, in November 2020, socially distanced house viewings continued. There were a lot of them, following Sunak's decision as Chancellor in the summer of that year to temporarily exempt the first £500,000 of the price of a residential property from stamp duty, up from £125,000, at a cost in lost tax of £3.8bn. Sunak deliberately and consciously inflated prices. The tax saving was around £2,000 on the purchase of a property at the UK average house price at the time. Some saved as much as £15,000. Those reaping the rewards included buy-to-let landlords and purchasers of second and third homes. Partly as a result of this and partly due to increased demand for bigger properties with gardens after lockdown, property prices surged 20.4% in the three years after Covid-19 arrived, compared to 7.8% in the three years prior to the onset of the pandemic, according to the Halifax. In cash terms, the average UK house price jumped by £48,620 between the start of 2020 and the end of 2022.* So many people rushed

---

* 'Three years on: how the pandemic reshaped the UK housing market', Lloyds Banking Group, February 2023

to purchase at that time, brokers and lenders reported gridlock. When Kish remembers these weeks, his blood pressure rises. Covid was delaying his wedding and he was one of those trying to buy a first home. The broker tried to convince him to fix for just two years. Miraculously, he ignored the advice and set for longer.

For Kish and millions of others, the political and economic impact of falling house prices is a live issue. According to the Nationwide house price index, there were double digit rises just twelve months ago. The time of writing is the sixth successive month of falling prices. It won't rebound soon. In June 2023 I interviewed the former Governor of the Bank of England Mark Carney. He does not expect to see a return to ultra-low interest rates for years, if ever, and says mortgage holders should enjoy whatever time they have left on their fixed rate deals before the new reality bites them.

The long-term impact on house prices of sustained mortgage rates of 5–6% or higher will be profound. Carney's former Bank of England colleagues stress-test the UK economy annually, seeking to get a sense of its resilience against shocks. Their scenarios are reasonable worst cases, not forecasts. In their most recent 2022 test, the Bank sought to look at the consequences of a UK residential house price fall of 31%. For comparison, even in the 2008 banking crisis, prices fell by much less, 17%.* The Bank says the 31% is 'at the tail of the historical distribution, and broadly comparable with a number of past severe housing market downturns in other advanced economies'. But some elements of the test are coming true. In 2022 they modelled the impact of 6%

---

* 'Stress testing the UK banking system: key elements of the 2022/23 annual cyclical scenario', The Bank of England, September 2022

interest rates, and Bank Rate at the time of writing is 5.25% and rising. Some mortgage rates are already over 6%. The UK is living through a real-life stress test.

So, is a 31% fall in house prices outlandish? In 2022 Andrew Goodwin of Oxford Economics suggested the affordability of property relative to people's income levels indicated the market was overpriced by about a third. His research group looked into trends in global house prices after busts that had followed booms. They found that when prices surge – as they have done in recent years – around half of the gains are lost when the market slumps.* In individual examples, such as the UK in the 1970s and 1990s, the slump was closer to 100% of what was gained during the boom. They note that, since 2012, global prices have risen about 40% and it is highly likely that some of that is about to be wiped out. Oxford Economics' core expectation, however, is not for a massive 30% correction in one swing. They predict that the UK will experience moderate sustained falls, but crucially they see no recovery until 2025, substantially longer than in other comparable nations where inflation is more under control.

The prediction that there is little to no chance of prices being on the up again by the time of the next election is of concern to a governing party with a traditional owner-occupier voting base. In fact, the most compelling explanation of the underlying driver of why so many self-identifying older working-class voters backed the Tories in former Labour 'red wall' seats in the 2019 general election is that most of them had acquired their own homes, many thanks to Thatcher's 'right to buy'. The economic profile of voters in the red wall is more similar to traditional Conservative voters than many have assumed.

---

* 'Research Briefing', Oxford Economics, January 2023

At the moment, the government is trying to lessen the pain for the most exposed homeowners rather than eliminate it. In response to pressure from the consumer finance expert Martin Lewis, the Chancellor has agreed a so-called charter with the big banks, by which they commit to switch struggling borrowers to interest-only loans and not to repossess for at least a year after a borrower fails to keep up the payments. Labour has promised a slightly more binding version of the same. It's a dampening of the crisis, not an elimination of it. There are concerns renters are still at the sharp end, because buy-to-let landlords won't be helped by the charter, and there is a risk rents will continue to go up significantly.

If the government holds its nerve and doesn't resort to one of its traditional bungs to homeowners, that would be a good thing. A reverse wealth effect, when consumers stop spending because they see the equity in their houses shrinking, is always scary for a government. But building prosperity on the basis of the illusory wealth trapped in our homes means the UK economy is always on fragile foundations – and it foments generational fracture, with so much of the wealth accruing to older people, and the young priced out of the market.

★ ★ ★

But the Great Transfer will surely end the injustices of a world in which those over fifty own everything. This is the idea that Britain, like much of the West, is on the cusp of a £5.5 trillion transfer of wealth from Boomers to Millennials over the next thirty years.* There is a problem though. While reducing inequality between generations, it will increase

---

* 'Wealth Transfer in the UK', Kings Court Trust and CEBR

inequality within each generation, because inheritance is such a driver of privilege. Research from Oxford University in 2020 found that 84% of those gaining wealth transfers like inheritances own their own home. Their average wealth held was £500,000. For those without such bequests, only 60% were homeowners and their average holdings were £220,000.* This unfairness is set to get much worse. As the IFS and Nuffield Foundation explain:

> *Inheritances are likely to become considerably larger, not just in absolute terms, but also relative to lifetime employment income.*†

Their projections found that for one in ten of those born in the 1980s (the youngest Millennial cohort they analyse), the inheritance they are likely to receive will be more than half of the average lifetime earnings for someone of their generation. In a single moment of inheritance, those individuals will be rewarded with what most in their age group would only be able to acquire through decades of work. Another manifestation of how inter-generational unfairness is morphing into intra-generational unfairness is that 54% of the top fifth of earners received a gift in the last eight years, according to the IFS, whereas just 13% of the bottom fifth were beneficiaries. As for inheritances, the scale of what people are in line to receive is very unevenly distributed. One in five people born in the 1980s are likely to inherit

---

* Nolan et al., The Wealth of Families: The Intergenerational Transmission of Wealth in Britain in Comparative Perspective, Oxford University, August 2020

† 'Inheritances and inequality within generations', IFS and Nuffield Trust, July 2020

less than £10,000, but the top 10% will inherit on average just under £500k. Graduates born in the 1980s are expected to inherit almost twice as much as those whose highest qualification is at GCSE level. Neither gifts nor inheritance are a level playing field. It goes against the grain of 'levelling up' too. Older parents in London own twice the wealth of those in the North East.

Even for those bailed out financially by the death of grandparents and parents, the windfall is often too late to make the kind of difference they would ideally want. We are living longer and that means people come to inherit much later. The average age of people losing their last surviving parent is forecast to be sixty-four for those born in the 1980s. For a third of people in that age cohort, this will not happen until they are in their seventies – which is far too late for them to provide even their children with the secure home they would have wanted.

Inheritance does often leapfrog a generation. Money passed down from grandparents to parents bolsters the reserves of the already substantial Bank of Mum and Dad, or the £7.4bn a year transferred from parents to children in gifts. As the *Sunday Times* Data Editor Tom Calver notes: 'if the family were a real bank (where that money actually had to be paid back), it would be the tenth-largest lender in Britain'. But it is still the case that 90% of wealth is handed down through estates after death rather than earlier.

There are some who argue that if only social care were less of a lottery, if there was a proper safety net, then parents would be more confident about making gifts of life-changing sums earlier. But this is just another way of saying that the wealth in the older generation's housing stock is not as substantial as the headline numbers would suggest – because

it is to make the dubious case that we should not take personal responsibility for the costs of our infirmity in old age and should instead transfer them to the state and young people. If inheritance tax was set at close to 100%, then there is a case for saying that the state should pick up all social-care costs. I don't however hear a clamour for the state to prevent all wealth cascading between the generations.

In fact, the political pressure is in the other direction. A number of senior Tory MPs, including Nadhim Zahawi, Priti Patel and Jacob Rees-Mogg, want inheritance tax abolished altogether. The economics of their case is puzzling, but the politics is simple. A recent Ipsos poll asked the public which taxes they considered to be most unfair, and inheritance tax topped the bill with 43% naming it. It's not an untypical finding.

As an example of how we make judgements about tax based on who we want to be, rather than who we are, it is relevant that the public seemingly hate a tax that won't apply to most of them. Fewer than one in twenty-five deaths lead to inheritance tax being levied by HMRC. There is a headline tax-free threshold of £325,000, but with additional allowances for transferring a main home to children or grandchildren it can be magnified to £500,000. Because couples can combine their individual allowances, it is possible to bequeath £1m without incurring a penny of tax. That said, it's not a completely trivial tax for the Exchequer. In 2010 inheritance tax brought in £2.5bn, but with house prices soaring it is on track to yield more than £7bn a year. Which rational government will give that up when public services are failing for want of cash? But would a rational government increase the rate of inheritance tax? Political history makes that unlikely. Back in the autumn of 2007, a pledge by the then leader of the opposition, David Cameron, to restructure inheritance tax so almost no one would pay it appeared to be popular – and deterred the then prime minister, Gordon Brown, from holding an early general election. More recently, Theresa May's 'dementia tax' – her proposal that older people should use more of their accumulated savings to pay for their care – was widely seen as having been the big reason her party lost its majority in the 2017 general election. Trying to deprive Baby Boomers of their right to transfer their savings to their children – or their pets – does not appear to be a vote winner.

According to research by the think tank Demos*, 55% of people believe inheritance should always be completely tax-free. And people objected to it even when they didn't expect it to affect their own inheritances or bequests. They said it wasn't fair for tax to be paid on savings that had already been taxed. Zahawi makes this argument in his essay calling for its abolition. But most of the capital gains in our houses are gains that have accrued to us largely because of when we were born, rather than because of any great talent we possess, and it is hard to see them as having any great link to our taxed income. As it happens, when Demos asked more detailed questions, it found the majority only oppose inheritance tax on savings accumulated from taxed wages or personal items like furniture. And when asked what inherited sum should always be tax-free, the average they identified was £300,000 – or well below the actual tax-free threshold.

If politicians want to look for tax reforms in this area, there are other solecisms in the system as it applies to our homes

---

* 'The Inheritance Tax Puzzle: Challenging assumptions about public attitudes to inheritance', Demos, June 2023

worthy of their attention. It's bonkers and regressive that £40bn in council tax is levied from a system of bands supposedly related to the quality of our individual houses based on selling prices in 1991. Because some neighbourhoods have become much more desirable since then, millionaires, even billionaires, find themselves in lavish homes that are not in the highest council tax band. This could and should be fixed. As for stamp duty, it is a terrible tax, because it makes moving house much more expensive, when any sensible government would want to encourage as much mobility as possible – to encourage empty-nest older people to move out of inappropriately large properties, and to make it cheaper for people to cross the country for a new job. Paul Johnson of the IFS perhaps sums up best:

> *Of all the taxes levied at present, Stamp Duty Land Tax – the tax you pay on purchasing a property – has a pretty good claim to be the most damaging and pernicious of the lot. The more often you move, the more tax you pay. It gums up the housing market and, by extension, the labour market. Mutually beneficial transactions, for example an older person in a big house trading places with a younger family in a smaller house, are disincentivised.**

Rather than total abolition, the Resolution Foundation suggests halving stamp duty rates and making permanent a £250,000 tax-free threshold that is due to fall to £125,000 in 2025. At a cost of £3bn, it won't happen unless funds can be raised elsewhere. But it makes sense.

---

* Paul Johnson, 'Stamping on stamp duty would free empty nesters to fly their coops', Institute for Fiscal Studies, January 2023

Changes to the tax system alone won't fix our current housing malaise. The prominence of inheritance tax in political debate points to an important change in how we see ourselves. Britain used to tell itself a story about owning a home as the deserved prize for doing the right thing, studying hard, getting a good job and reaping the rewards. Many now view that as fiction. Instead, parents and grandparents believe that their descendants' only hope of acquiring a home is via the transfer of the wealth locked up in their properties. Which is partly why so many think inheritance tax is wrong. But this is to ignore that the older generation's housing wealth is just another symptom of the dysfunction that stops young people climbing on the ladder. Inheritance tax is therefore just one part of a redistributive tax system that could yield a solution, if the tax revenues were deployed to get Britain building again. Inheritance tax needs to be reformed not abolished – so that it raises more money in a fairer way. This would involve closing loopholes, abolishing exemptions, and then changing it from a flat rate 40% levy to a progressive tax, with bands going from 20% up in 10% steps to a top rate of 50% on the very largest estates.

We should challenge ourselves to change the hoarding instinct we all have as we grow older to one of altruism. Is there a way to create a spirit of civic duty to deploy some of our accumulated property wealth to help younger people more generally? This issue arose when a pair of respected economists, Olivier Blanchard – formerly chief economist at the IMF – and Nobel prize-winner Jean Tirole, were asked in 2020 by the French president, Emmanuel Macron, to address the big economic problems. In their work on demographic shifts and widening inequality, they suggested that the receipts of taxes on estates could be used explicitly and only for articulated social purposes. Economists usually think of such 'hypothecated taxation' as a confidence trick, because all our tax payments go into one big pot and the money itself has no idea what it is being spent on. But Blanchard and Tirole suggested the proceeds of inheritance tax could be used to create 'individual accounts that the disadvantaged young could spend to avoid having to work while studying or training, or financial accounts that disadvantaged kids could access when becoming adults'. Harnessing the receipts to improve the life chances of children and young people is a powerful idea. It should be at least as resonant for most of us as leaving assets to our own offspring. If targeted well, it could help disadvantaged young people to acquire skills and qualifications currently beyond reach or help them set up businesses. It would turn the payment of inheritance tax from an annoyance to a matter of personal pride. And maybe it could also fund interest-free lifetime loans, to provide a deposit for that vital first rung on the housing ladder.

CHAPTER 9
LEARNING HOW TO LOSE

Friday, 9 June, 8.10pm. I am sitting on the stage of the Adam Smith Theatre in Kirkcaldy, the *Lang Toun*, or Long Town, in Scotland's Kingdom of Fife. It is the 300th anniversary of the birth of Kirkcaldy's most famous child, Adam Smith, who is to economics what Newton was to physics. They were extraordinary, almost miraculous, pioneering thinkers, who changed how we think of our world in a basic way, who gave us the tools to navigate and shape it. And even though Smith's 'rules' are social, about the relations between people, and therefore less certain and predictable than Newton's, his insights remain relevant and important. He helped us understand how competition between firms incentivises them to cut costs and prices, how specialisation boosts efficiency and productivity, how lower prices amplify the spending power and wealth of consumers, and how efficient markets require clear and fairly enforced rules overseen by an uncorrupt government. These precepts still set the parameters for how to preserve and enhance our prosperity. When they've been systematically ignored – as has been the case in liberal democracies but is more often associated with the arrogance of totalitarian states of left and right – the consequences have usually been ruinous.

His reflections on the psychology of business owners are

apt. At the birth of capitalism, he put us on notice that business owners will seek to maximise profits no matter what the detriment to customers' and employees' living standards. Proprietors will simultaneously complain that workers are charging too much for their labour while setting the prices of their own goods and services at levels that may be impossibly and painfully high for some customers. The implication was – and is – that we must always be watchful of whether there is true and effective competition in any particular market, because when there isn't we'll invariably be paying too much for important goods and services, and employees may well be receiving too little. While recognising that markets are the best distributional system of which we know – arguably the most powerful information-processing neural network ever created, until our new age of artificial intelligence and quantum computing – we should trust their price-setting mechanism warily, and be prepared to intervene where outcomes are unjust or suboptimal. Smith's 'invisible hand' is a flawed god. Almost uniquely, therefore, Smith is a hero of right and left. He is guru to Margaret Thatcher and to Gordon Brown, the former prime minister who invited me to stand in front of the lectern and mark Smith's tercentenary.

This place, Kirkcaldy, is a crucible of so many of the forces that have remade modern Britain. Brown stood down as the MP for Kirkcaldy and Cowdenbeath at the general election of 2015, having represented the area since 1983 (before boundary changes of 2005, it was called Dunfermline East) and its current MP, Neale Hanvey, is a member of the Alba Party, the Scottish nationalist and pro-independence party that broke away from the nation's ruling SNP. Brown grew up here, as a 'son of the manse' or child of a Church of Scotland minister, who delivered his sermons in a battleship of a Victorian church, St Bryce Kirk. At its peak, St Bryce seated 1,150 celebrants. Today, the congregation is a fraction of that. It's a town of solid, grey-stone, nineteenth-century buildings that could do with a scrub and a touch-up. It speaks to an earlier and lost time of economic and intellectual confidence. Brown himself lives down the road. His study looks over the expansive Firth of Forth, which is spanned by the magnificent cantilevered Victorian railway bridge, and which a century ago was congested with barges and ships transporting coal from Fife and Lothian to the rest of Britain.

The Adam Smith Theatre, originally known as the Adam Smith Halls and opened in 1899 by the Scottish American steel magnate Andrew Carnegie, has the size and proportions of a stately home. It is inconceivable that a public entertainment building of such size and grandeur would be constructed in such a relatively small British community today. It's the first time it has been in use after months of redecoration and refurbishment, and I feel both honoured and a fraud as the first performer, as it were. It is not going to be a night of Adam Smith exegesis or rip-roaring fun.

For an hour, in front of the kind of audience I love – engaged and polite, not the tribe who shout at me on social media – I have been banging on about national and global risks. It's been an attempt to explain the implications for our way of life of the return of inflation, the end of free money, the proliferation of world-changing generative artificial intelligence, the recasting of China from the engine of global growth to perhaps the most serious threat to our way of life, the long game to defeat Putin, the potentially tragic failure to stay on track to limit global warming to 1.5 degrees above pre-industrial temperatures. I make my point about the massive information processing power of Smith's invisible hand in the marketplace and the mostly rational sorting of information by generative AI's large language models. I don't claim originality for the thought, but the part of the metaphor that I stress is the analogous flaws in both GPT-4 and the market – how the market and AI services like GPT-4 often are subject to errors, what are known as 'hallucinations' in the digital jargon. This doesn't make them redundant; it simply means they must always be subject to human judgement and correction. My aim in describing how the UK and much of the West is not working for millions of people – how so many are condemned to low and stagnating incomes, permanent struggles to pay bills, poverty of ambition – is not to encourage defeatism or a sense of hopelessness. It's to show that opportunity is the obverse of failure; it is to reveal an enemy in order to defeat it.

I move to a direct conversation with the audience, questions. There is time for just one more. A man in the middle of the auditorium is given the microphone and starts by informing us that he'd just seen a breaking story on Twitter: Boris Johnson has announced he would be standing down with immediate effect as an MP. It's an electric shock through the auditorium. Johnson may no longer be prime minister. But, for better or worse, he's the most important politician of his generation. His departure is the end of something important. For *this* audience, it's good news, especially after my Old Testament sermon. There is spontaneous applause, which gives me a moment to read my phone. I was already aware that Johnson had been sent the draft of a report by the Privileges Committee, the de facto MPs' court, into whether he had committed what are known as 'contempts' of Parliament by lying to MPs about the parties that took place in Downing Street in breach of Covid-19 restrictions. Clicking on WhatsApp, I now read Johnson's confirmation of the Committee's draft conclusions, namely that he recklessly lied in the House of Commons when telling MPs there were no parties and that there was adherence to all those rules and guidelines that he himself had written and announced. Johnson is incandescent with the unfairness of it all. He's a stitched-up kipper, he insists, the innocent victim of the roughest of justice. A conspiracy of the Remainer elite has finally done him in. He is a martyr to the noble Brexit cause, and as such he can't possibly recognise the authority of the court. They can shove their punishment – suspension from Parliament and a referendum of his constituents on whether his licence to represent them needs to be renewed in a by-election – up their own self-righteous posteriors. He's off to kinder and more remunerative pastures, where he can make millions writing and talking about his favourite subject, himself. And sod it that he's supposed to keep the report confidential till the Committee publishes. Such niceties apply to smaller men.

Johnson's departure was of a piece with his life, the melodramatic end of perhaps the penultimate episode of an imaginary Netflix series, *Just Boris*, or *Tory Succession*. Flashback to 1982. His father, Stanley – bottom-slapping (which SJ doesn't recall), Brussels mandarin – reads a letter written by a master at Eton College, the most famous boys' school in Britain, perhaps the whole world, the training ground of the British ruling class, educator of twenty prime ministers out of fifty-five since Walpole in the eighteenth century, and of a generation of contemporary ministers. 'Boris sometimes seems affronted when criticised for what amounts to a gross failure of responsibility,' writes the Eton master Martin Hammond of seventeen-year-old Johnson, in his reference for Johnson's application to read 'Greats' (classics) at Balliol College Oxford. 'I think he honestly believes it is churlish of us not to regard him as an exception, one who should be free of the network of obligation that binds everyone else.' Flash-forward thirty-one years. Harriet Harman, 'mother' of the Commons, alumna of another manufacturer of the UK's self-assured leaders, St Paul's School for Girls, is working late into the night at a utilitarian desk, illuminated by an unflashy anglepoise lamp. As chair of the Privileges Committee that has been adjudicating on Johnson's lies, she is drafting the judgment. She types: 'we came to the view that some of Mr Johnson's denials and explanations were so disingenuous that they were by their very nature deliberate attempts to mislead the Committee and the House, while others demonstrated deliberation because of the frequency with which he closed his mind to the truth.' She allows herself a contented half smile and closes the laptop. Johnson's fate is sealed.

It's all been such theatre. When I first saw Boris Johnson at the Dispatch Box of the House of Commons on 25 July 2019 on his debut in the chamber as prime minister, it did not feel real. Part of my detachment stemmed from the challenging idea that someone I thought of as a journalist could end up in that job. It wasn't that I felt 'it could have been me'. It was that I was certain that I am not qualified to do that job, and Johnson always struck me as even less qualified. And yet here he was.

For much of my conscious working life, he'd always been around. I knew his siblings, Jo and Rachel, better, partly because they overlapped with me when I worked at the *Financial Times*. His and my path crossed periodically, though, at London parties, EU summits and via mutual friends. There was one summit, in Cardiff I think, when we were both at a loose end and went for a long walk along the waterfront. All he seemed motivated to discuss was how much money was being made by mutual contemporaries of ours. He did not hide his envy. Funnily enough, about ten years later, I had an almost identical conversation with George Osborne, who was obsessed with the enormous sums accumulated in the hedge fund industry by his friends from Oxford University. On any normal yardstick, Johnson and Osborne were doing well for themselves, so their annoyance about others in their circle becoming so rich has stayed with me.

A few years later, in 2001, I briefly worked for Johnson. I had a very part-time job as a columnist and associate editor of the *Spectator* magazine when he was editor. It was, of all my jobs for newspapers and TV, the most chaotic, because he could not make up his mind what he wanted. We would talk a few days before deadline and agree a theme, and then invariably he would ring hours before the paper was due to go to press and ask for something different. Resigning from the paper was a liberation. After that, when he became London Mayor, I ran into him less. I once gave a speech in honour of his former father-in-law, the BBC's distinguished former foreign correspondent Charles Wheeler, and was charmed by his then wife, the barrister Marina Wheeler, and their children. He turned up after I'd spoken, in time for the drinks. On another occasion, he arrived midway through the main course at a private dinner at the World Economic Forum, hosted by David Cameron and laid on by a prominent party donor. I was there with a couple of other journalists, Cameron's aides and George Osborne. Johnson arrived with his mayor's entourage. There was a mild kerfuffle about whether there was room for them. There wasn't. I experienced that same sense of detachment, like watching a movie, that I felt when Johnson became prime minister. Back then, Cameron and Osborne hadn't long been PM and Chancellor.

Here were these three relatively young men, two of them from the same school, Eton, the other from another privileged fee-paying school, St Paul's. They'd all known each other for years, and had grown up with each other, to an extent. They spoke and acted as though it was the most natural thing in the world that they were in charge of the UK and of its most powerful city. It was the mutual contempt that they manifested for each other, the joshing and bantering, that was particularly disconcerting. It was as though they were all in on a private joke, from which the rest of us were excluded. I'd like to think the joke was that they realised how extraordinarily fortunate they were, but I don't think it was that. They gave not the slightest hint of doubt about their entitlement to run the country, though then and now I struggle to know what they thought the point of it all was.

Between Cameron and Johnson, Cameron was always the more straightforward. At another World Economic Forum meeting, there was a small party of UK officials and hacks in a chalet. It was well after midnight. Cameron and Osborne were there and drinking with the rest of us. For them, however, it was a waiting room. They were expecting to be granted an audience in the early hours of the morning with Rupert Murdoch, who was presumably still functioning on the US time zone. I asked Cameron why he demeaned himself by fitting in with Murdoch's exhausting schedule rather than vice versa; why he was allowing himself to be cast as courtier rather than king. He said, matter-of-factly, that was just the reality of power politics. The point about Cameron is that he was much less confusing as a prime minister than Johnson. Even when I thought Cameron was making serious policy mistakes – holding the EU referendum, the scale and nature of public spending cuts, of austerity, and especially Osborne's savage cuts in capital spending – I understood his motivation. Quite a lot of Johnson's time in 10 Downing Street was simply baffling.

I still feel slightly traumatised by the experience of reporting on Johnson's three years as prime minister, because his genius is to create the kind of chaos that constantly distracts from a proper evaluation of what the government is actually doing or not doing. He championed Brexit as restoring the power and authority of Parliament, and then illegally sent MPs home when they wanted their say on the terms of leaving the EU. He pledged his Brexit deal was not putting a border in the Irish Sea between Britain and Northern Ireland, when everyone could see that it was, and then subsequently railed against the border that he'd created. He told us all to imprison ourselves in our homes during the pandemic and warned us that any socialising was imperilling the entire community, and then he lived above party central in 10 Downing Street. He branded himself as the prime minister who would 'level up' and abolish the cancer of inequality in Britain, and then devoted huge time and effort to raising well over £100,000 from friends and admirers to lavishly redecorate his new Downing Street home (and ended up taking out a loan to pay for it).

There is an incident that captures how confusing it was to work for him. It is just one of many that his former colleagues have regaled me with, often shuddering in the telling. The setting was the regular Downing Street morning meeting held to review media coverage. James Slack, who has been both official spokesman for the PM and director of communications, pointed out to assembled colleagues, including the PM, that a story on the front page of the *Daily Telegraph* was wrong and he would make sure the paper corrected it. 'Ah yes,' said the PM, 'I might have spoken to Chris Evans [the *Telegraph* editor] last night.' One of his aides said to me: 'One of the nightmares for Lee [Cain, the PM's director of communications 2019–20] was Boris was running his own media operation with Carrie [his wife] from the flat [where he lived]. Everyone knew downstairs that lots of leaks and briefings were either him or her, with him on the sofa next to her.' It was almost transparent. There was a period when the morning political email from the Politico website was like an unofficial press release from the Downing Street flat. Johnson, however, was obsessed with newspapers above all other media. He was especially close to the proprietor of the *Standard*, Evgeny Lebedev, whom he ennobled, the *Telegraph*, whose owners employed him for decades, and the *Daily Mail* – whose daily paper has backed him through thick and thin, and whose influential former editor, Paul Dacre, Johnson tried to put in the Lords. In March 2021, Rebekah Brooks, who runs all Murdoch's newspapers and other media in the UK, recruited Johnson's then director of communications, James Slack, to be deputy editor in chief of the *Sun* newspaper.

An official told me that Johnson's morning meeting with aides was 'all about [what was in] the papers' when 'it should be 95% about policy' – and this despite the Tories' own research that showed voters got '60 per cent of their information from television, 25 per cent from digital and the rest from newspapers'. Another said: 'Johnson acted more like a newspaper editor than like a prime minister. He ran the morning meeting like a dysfunctional shambolic editorial meeting, going through what was in the papers.' The official added: 'it goes without saying he was late for almost all of them, despite having the shortest distance to travel, and was always with wet hair/untucked shirt/poorly shaved, and in desperate need of a pot of coffee.'

Here is an account of those morning meetings from someone in the room:

The set-up changed once Covid kicked in. Before Covid, these meetings took place in the PM's study. There is a large round table that can seat about eight people and Boris would sit there, always with his back to the Cabinet Room – where a private secretary would sit – and would face the maximum number of people, with a clear view of the door. His PPS [principal private secretary], the Cabinet Secretary, James Slack [at the time his official spokesman], Ed Lister [a senior aide], the director of communications, the chief whip and Chancellor would sit at that table, whilst other advisers and private secretaries would sit around on the sofa and chairs. During and after Covid, these meetings would take place in the Cabinet Room. The PM would sit in the centre of the table in the only chair with arms and people would spread around the table or join on Zoom. The cast list generally included Matt Hancock, Chris Whitty and Patrick Vallance too.

Dominic Cummings tried hard to prevent the meeting being all about the press and media coverage, according to a civil servant. He would meet with officials and special advisers before the meeting to establish what government decisions actually had to be taken in the meeting. But:

The PM liked to start the meeting in exactly the same way: James Slack or Jack Doyle [a subsequent director of communications] giving an assessment of that morning's news. This would be a prepared printout, rather than bringing in a stack of newspapers – though Boris would regularly flick through the papers at a small table outside his office over the course of the day. Boris liked discussing which outlet was saying what. Occasionally he could be very worked up, and would arrive having got himself into a lather about a certain piece. An ongoing problem was his desire to respond to a story or to create news or generally shape the news cycle, rather than set the agenda for his team for the day. Unless an issue was an active one – such as a Covid issue, free school meals, a tricky vote in Parliament – he rarely asked for an update on a conversation we'd had the day before, because the news was, well, new. Inevitably it was Cummings who gave everyone in that meeting a sense of what was needed from them for the following twenty-four hours and would pointedly ask difficult questions of people.

Johnson's notion of government as a news-generating machine infected everything. 'It influenced how his own policy advisers presented their papers to him,' an adviser said. 'They would start them with a pretend dream headline, or a quote from Churchill or Gandhi.'

More than any prime minister in my experience, he was constantly talking to a coterie of newspaper editors and executives to whom he felt close. One editor told me that the close relationship Johnson had with the proprietor distorted all their coverage. An official said one reason he fell out of favour with Johnson was that the editor of the *Daily Telegraph*, Evans, had complained to the prime minister that his paper 'wasn't getting enough stories'. It cuts the other way. When Johnson read stories in newspapers that were damaging to him – and there were loads, given his propensity to see a rule of propriety or ethics and decide it didn't apply to him – he would routinely clunk around Downing Street calling for the relevant journalist to be sacked. 'He would rant and rave about stories he didn't like. He was particularly incensed when Carrie was upset by something in the papers.'

He told Dominic Cummings that he was trying to get Rupert Murdoch to fire the editor of the *Sunday Times*, Martin Ivens, after Ivens ran a series of stories about whether as London Mayor he had given improper help to Jennifer Arcuri, an American tech entrepreneur who says he had an affair with her. Coincidentally or not, Ivens was replaced as *Sunday Times* editor in January 2020 but joined the board of Times Newspapers. A different aide heard Johnson on the phone talking to a newspaper executive calling for another journalist to be fired over a report that was embarrassing for him. In total, I've spoken to six senior journalists, all responsible for critical coverage of Johnson, who were sacked or moved sideways after Johnson complained about them. Colleagues at their organisations and in Downing Street corroborate that Johnson personally intervened when he was offended by a report. It's impossible to prove what is in the head of a proprietor, executive or editor when they change the responsibilities of a colleague who has enraged the prime minister. But it is never easy for the boss of any newspaper organisation to have the most powerful person in the country harrumphing on the line, especially when he is showering favours on rivals. In 2019 and especially after Johnson won that eighty-seat majority at the end of that year, newspaper executives and owners believed he was going to be the most powerful person in the country for a decade at least. They wanted exclusive interviews with him. They wanted the inside track on what he was doing. They wanted any help he could provide when their businesses were facing pressures from digital media that could wipe them out. They did not want to be permanently on the wrong side of him. Johnson knew this and exploited it. The risk for the country of course is that when newspapers become co-opted into a prime minister's court, they abandon the responsibility of holding them to account and to the promises they've made to the country.

Perhaps a small example from my own recent life is instructive. At 4.48am on 13 April 2022 – when I was asleep – a WhatsApp was sent to me by Johnson's director of communications, Guto Harri. He cut and pasted a chunk of a tweet that I'd written the previous day. It was about how the police had fined Johnson and the Chancellor Rishi Sunak for gathering in the Cabinet room to celebrate Johnson's birthday, in breach of the Covid-19 social distancing rules he'd written. The fine was prima facie evidence that Johnson had knowingly lied to MPs when he told them there had been no parties in Downing Street and all Covid-19 rules had been followed. Harri focussed on the conclusion of my tweet:

> *If Tory MPs unthinkingly keep him in office without a proper and public assessment of how parliament was misled, because that is what suits them, and if they blithely ignore the Ministerial Code, then the charge will stick that this or any party with a big majority is simply an elected dictatorship, and the constitution means little or nothing. This is not just a slippery slope. It is the bottom of the slope.*

What I was arguing is that the police verdict on Johnson meant MPs should investigate whether he as prime minister had deliberately lied to them from the Dispatch Box. And if they backed off doing that, it would show that the normal rules of probity could be overridden by a government with a big enough majority. This was not to prejudge whether or not Johnson was guilty. It was simply to say that there was a case to answer and it should be answered. Harri, in the early hours of the morning, took a different view. Above the copy of my tweet, he wrote 'Seriously???' And he fired off complaints about me to various bosses at *ITV News* – who respectfully disagreed with him that I had crossed a line or that editorial impartiality rules had been breached.

I naively assumed that would be the end of it. But there was then a story in the *Daily Mail* – one of the newspapers regarded by Johnson as an ally, for which he now writes an exceptionally well-paid column – under the headline 'Bias storm as Peston compares the PM to "an elected dictator"'. The report began:

> Robert Peston was accused of breaching ITV's duty for political impartiality last night after he criticised Boris Johnson for not resigning over Partygate. ITV's political editor tweeted that if Tory MPs decided to keep the Prime Minister in office, it would look like Britain was 'an elected dictatorship.' And he added: 'this is not just a slippery slope. It is the bottom of the slope'. Senior government sources said that such comments would not be allowed to be conveyed on air because it would fall foul of the duty of broadcasters to be impartial. They said Mr Peston should be held to account by his employers, even though the comments were made on social media.

I read this and wondered about the identity of 'senior government sources' and what they thought they were doing. The tweet didn't say what they claimed. I hadn't criticised Johnson for not resigning and I hadn't said Tory MPs had to remove him from office. What I did argue – and what actually happened, when the Commons Privileges committee subsequently decided to examine his statements to MPs about the parties in Downing Street – was that MPs should investigate whether he had knowingly and wilfully misled them. I was confident there was nothing I said in that tweet that I wasn't entitled to say on air, on *ITV's News at Ten* and on its *Evening News*, under the impartiality rules overseen by the regulator Ofcom. I was not gaming the laxer regulation around social media to make a point I wouldn't broadcast on television. In fact, on the night that I published the tweet, which was also published as a blog on ITV's website, I arguably went just as far in my televised conversation with Tom Bradby outside 10 Downing Street: I said 'it was a big, and many would say, really damaging moment for the British constitution, for those things we pride in our democracy' that the prime minister was not explaining how exactly he came to mislead Parliament and was not contemplating resignation. No one, as far as I know, complained about that broadcast.

For several days, *Mail* journalists rang my bosses asking if they were going to apologise for what I had written, if I was going to apologise or if I was going to be punished. None of that happened, because we were sure we were just doing our jobs, of holding power to account. But it felt like bullying, coordinated by those 'Downing Street sources'. When there is a furore of this sort, significant amounts of my and my colleagues' time is captured by our need to review what I've said and done. It's tiring, distracts from other work and is demoralising. The chaos of Johnson's 10 Downing Street was exported to us.

It was typical that I should be on a theatre stage in Kirkcaldy when this chaotic era should be ending. After assessing Johnson's statement on the still-unpublished verdict of the Privileges Committee, I am asked to comment on him packing in his entire parliamentary career. 'Oh fuck,' is all I can manage. My questioner perseveres. 'But what does it mean?'

Me: 'It means I'm in the wrong fucking place.'

Cue riotous laughter. At my predicament in being five hours from our television studio in central London. Maybe there is reluctant admiration at Johnson's skill in always being the centre of attention. Johnson as prime minister was never popular in Scotland and was the recruiting sergeant for the Scottish National Party's campaign to win independence from Westminster. But no one doubted his significance in the life of the nation. We were together for one of those symbolic turning points, but where would we end up?

There were worse places to consider the meaning of Johnson's end. Kirkcaldy's modern story is the precise tale of decline that his Brexit boosterism was supposed to reverse. It had been, in the UK's Victorian industrial heyday, a thriving manufacturing and coal mining town. It was a centre for making linoleum, the functional floor covering of a million kitchens in the back-to-back terraces and council houses of a prospering working class. Today, skilled, well-paid jobs are in too short supply and Kirkcaldy is one of Scotland's poorest towns. Deprivation is rife. HMRC and DWP figures from March 2022 revealed that 40% of children in the constituency are living in poverty. In the Linktown Central area of Kirkcaldy, three-quarters are in that desperate category. These statistics are the caption to lives blighted by drug addiction, marital abuse, relationship breakdown, hunger and cold.

There are also, in the midst of Kirkcaldy's despair, examples of hope and public service. One is a social enterprise called the Cottage Family Centre. It provides drop-in day-care facilities for families struggling with physical and mental health challenges, or just unable to eat and keep warm. Its playrooms and kitchens are light, well stocked and welcoming. Caseworkers wean spouses, often men, off heroin dependency, teach how to cook nourishing meals, counsel damaged children. And they want to instil hope of a better life. When I arrived, a re-energised mother and daughter had just returned from a few days at the seaside, staying in the charity's caravan. There's another associated project, sponsored by Brown, the Fife Big House. It's a 'multibank', like a foodbank only far more ambitious. It is a warehouse and distribution centre for everything from trainers through to blankets through to classy handbags, cleaning products and non-refrigerated food. The venture was started after my colleagues at *ITV News* revealed quite how much of Amazon's older stock was being sent to landfill. Every week it distributes, free of charge, boxes of essential goods to caseworkers, who have requested help for families in dire need. Brown told me he is urgently trying to organise the setting up of other multibanks. One of the caseworkers told me there are some families where the children simply never washed with soap or shampoo before the Big House was set up.

The Cottage and the Big House are inspiring institutions. But in a country as rich as ours, we shouldn't need them. It is shameful for the UK that the Office for National Statistics has been reporting that one in twenty people run out of food every fortnight. These are manifestations of a state in big trouble, of an economy that is not delivering fair outcomes, of the pernicious inequalities that Johnson said would be fixed by Brexit and by his programme of 'levelling up'. I was not in the wrong place as he headed for the exit. I was where the judgement on his time in office needed to be made.

At the start of my lecture, I had contrasted Johnson and Brown. I argued that Brown's contributions to the welfare of the UK – through the significant expansion of in-work benefits paid to lower income families, the depoliticisation of interest rate decisions by transferring them to the Bank of England, the massive expansion of funding for health and education – had been overlooked and underrated. Many of these achievements had been carried out by him when Chancellor. And as prime minister, his grip of the banking crisis – albeit a crisis that would have been less acute if the Labour government had better regulated the City – was influential and largely appropriate. That said, I have long argued Brown could have been more radical in breaking up the biggest banks and ending the discrimination of financial companies against the underprivileged.

Johnson, by contrast, has been less principled, less diligent, less benign and – in one way – more important. His mark on the UK compares with that of Thatcher, for better or worse. He, more than any other politician apart possibly from Nigel Farage, was responsible for the most important constitutional and economic change of the past half-century, leaving the European Union, both as leader of the official Brexit campaign and as the prime minister who signed the Brexit treaties. He entered Downing Street in the summer of 2019, after the chaos of Theresa May's attempt to execute the will of the people by negotiating the terms of divorce from the EU, and – under the influence of his senior aide Dominic Cummings – he chose to magnify and exploit the mayhem. It was purposive madness that culminated in a near landslide victory for his party in a general election, and then the formal moment of Brexit itself, followed a year later by a very basic free trade agreement with the EU.

He also matters as we've discussed for being lackadaisical as Covid-19 coursed from Wuhan through Asia to Europe, but then licensed Cummings to set up a taskforce that effectively developed, purchased and injected vaccines. The quintessential Johnson is the one who almost died of Covid, in full view of the nation, and then failed to stop the Downing Street parties during lockdown.

Johnson wove his contempt for proprieties – his perceived freedom from the obligations that bind the rest of us – into the fabric of our most important institutions, by lying to MPs, packing the House of Lords with cronies, giving gongs to hangers-on and cheerleaders. He took seriously the existential threats from Putin's invasion of Ukraine and from climate change. But he treated everything else as a game, an exercise in ducking, diving and dodging in the pursuit and retention of power. When he became PM in 2019, trust in the political class was at an all-time low. Rather than attempt to restore the reputation of Parliament and parliamentarians, he exploited punters' cynicism. With a cheery smile he governed on the premise that if venality is what the punters expected, he wouldn't let them down.

He also matters because he is the British incarnation of fake news, motivated by a desperate need for love and attention, a voracious appetite for money and power and an acute sense of victimhood. He has become reminiscent of Donald Trump, especially in the way he saw an elitist anti-Brexit conspiracy behind the punishment meted out to him by the Privileges Committee. Trump and Johnson owe much of their political prominence to channelling the chronic sense of unfairness held by a class of entitled dispossessed older white men, with whom they exist in a state of mutual interdependence. They show what can be achieved in politics and government by ignoring inconvenient facts and promoting their own truths – and this is before they had an opportunity to take advantage of the latest artificial intelligence programs that personalise and reshape information, images and videos to maximise their persuasive power.

Consider how Vote Leave's campaign, for which Johnson was the figurehead, converted crucial swing voters with two distorted ideas: first that the reallocation of the direct budget costs of EU membership could be redeployed to save the UK's National Health Service and second that Turkey would join the EU and prompt another great surge of migration here. One was a story of hope, which seems laughable today given that the NHS is in far worse shape than prior to Brexit. The other was a story of threat, equally absurd, as Turkey is further away than ever from being welcomed into the club of democratic European nations. If Brexit was built on distortions, it is appropriate that Johnson's time in parliamentary politics was terminated by Harriet Harman's committee calling time on his fictions.

All that said, Boris Johnson is perhaps the prime minister we deserved, because he reflected back to us so many of our own flaws. With his chuntering, flag-waving rallying cries about making Britain great again, he rode to power, via Brexit, on the back of a cancerous illusion. It is the idea of British exceptionalism, of imperial Britannia ruling the waves, of British England beating Germany in two world wars and one World Cup, of being home to the world's greatest playwright and pop group, and to Churchill, and to the last proper monarchy. It is a Britain that – as depicted in James Graham's play *Dear England* about the England football team and Gareth Southgate – is obsessed with reliving and reclaiming its victorious past but is paralysed by the fear of losing. In the play, Southgate and the team psychologist Pippa Grange agree that the route to success is through the players being honest with each other about their frailties and fears, and

– perhaps more than anything else – getting over the terror of defeat. They might just as well have been writing a prescription for a malaise that goes wider than the national sport. The important lesson is that winning is not everything, even if it is a lot. Progress matters more. Being good enough matters, as do recognising our faults, working hard and earning the right to prosperity and to be heard.

Brexit is a case in point. The slogan of the winning campaign was 'taking back control'. It resonated with millions of people who felt they and the country did not have enough control, over who and how many people come to live here and over the making of laws. It worked. But it rested on the idea that happiness and fulfilment require the UK to somehow be independently great again. It was based on the myth of plucky Britain, plucky Tommy, going it alone. But when so many problems are shared global problems – from climate change, to harnessing AI, to standing up to dangerous autocracies – the imperative is not to be the winner but to be good enough. We all need a sense of self-determination, of not being bossed around. But true contentment, and success, rests on knowing we can't control everything. Resilience requires us to know what about our lives is wholly in our own hands, and what requires cooperation, give and take, partnership. Extreme autonomy is not worth it if correlated with poverty and heightened vulnerability to disaster. Consider this in the context of your family. Whatever your role, you know that either imposing your will on the other members, or being imposed on, is the route to tension and misery. Sharing problems, listening and understanding are the requirements for contentment. The same is true at the level of the nation – which is not the same as the economic fallacy that household finances are a model for national finances.

This is to say that, with a degree of national humility, about what is possible and what's not, we can be positive and hopeful. At some point, it may lead us to want to rejoin the EU. But that is not the argument here. When I started this book, I feared it would seem defeatist, which is not in my character. I have always been a believer that we can shape our fate, though only if we first understand ourselves. So the theme of this book is that we must not give up on ambition but we must abandon the idea that the UK should always win and be the best. Instead, we should concentrate on being good enough. When we have institutions that are widely seen as world class, that should be a source of pride. 'Elitism' is not a dirty word if entry to the relevant elite is open to all equally, based on ability, irrespective of class, gender, sexuality, ethnicity, faith, and anything about us that is indelible. But simply repeating – as our leaders routinely do – that we have four of the top ten universities in the world, or that we have had significantly more Nobel prize-winners per head of population is a comforting mantra, not a route to renewed prosperity. It reinforces the idea that we do not have to work for our future, that there is something intrinsically special about us. It would be healthier to focus on how many more Nobel prizes we want to win, and think practically about how to do so, and how and whether to turn the award-winning research into practical outcomes, including businesses that can employ us and generate the tax revenues we desperately need.

Adam Smith matters because he gave us rules that when applied to our democracy help us to govern ourselves rationally for the common good. Boris Johnson matters because he shows us the dangers of seductive wishful thinking. I was glad to be in Kirkcaldy for his goodbye. There were few better places.

CHAPTER 10
ARTIFICIAL HOPE AND REAL HOPE

I've been writing books about the condition of Britain for twenty years, and observing this country as a journalist for almost twice that. This is the first time that I've smelled the rot of decline and fatalism. Too much has gone wrong too quickly: a Brexit vote that harmed us economically; Covid that harmed us physically, psychologically and economically; a government led by Boris Johnson that collapsed because its own ministers lost confidence in the probity of their leader; the return of inflation and higher interest rates; the threat from Russia and China; all adding to our longer-running flaws of low growth, too-acute inequality, too much deprivation and poverty. It's a lot.

This feels like a 1945 or a 1979 moment, when we need to seize our destiny in our hands and make big changes. But where are Margaret Thatcher and Keith Joseph or Clement Attlee and Aneurin Bevan when you need them? Where even are leaders of the competence and vision of Tony Blair and Gordon Brown? What's striking is that the policies of government and opposition are largely attempts to recalibrate the status quo, to patch and mend the existing system. The government's priorities are to help the Bank of England return inflation to target, to secure free trade deals with countries and regions outside of Europe that may partly compensate over decades for what we've lost in withdrawing from the EU's single market, to deter asylum seekers risking their lives by making dangerous Channel crossings in flimsy boats, to stitch up a wounded National Health Service. For Labour it's to invest £28bn a year in creating a greener economy, though only when resources allow, to 'reform' the NHS in ways that are not precisely specified, to lift the growth rate of the economy to make it top of the G7 group of rich countries, also in ways that have not been articulated in detail. You will have your own views about which of these promises and plans appeal to you, and which are alienating. But none of you, surely, will feel that what's on offer gets to the root of the matter, that there's a comprehensive programme to restore confidence that a fairer and more prosperous society is within our reach.

For both the Tories and Labour, there is a money problem – or rather the catch-22 of there not being enough money flowing into Treasury coffers to finance ambitious programmes of change, because growth in the economy is so low, but there not yet being a credible programme to revive growth that does not involve significant sums of money that don't exist. You might call it the Truss Paradox: she wanted to borrow £45bn a year to spur growth by cutting taxes; investors and markets said no. By the way, there are some of her supporters who defend her by saying that the interest rates set by markets rose in the spring and summer of this year to the penal rates they hit after her notorious budget, and therefore Jeremy Hunt's reversal of her tax cuts and imposition of relative prudence was all for nought. This is not just muddle-headed, it is dangerous. Goodness only knows where interest rates and inflation would now be in the absence of the U-turn. It is conceivable that the Treasury would have struggled to sell gilts, its debt, to investors. We'd have been in a full-scale sovereign debt crisis of a magnitude only Greece has experienced in modern times as a developed economy.

The bigger problem however with Truss's electric shock to the system is that it has discouraged political ambition, and has encouraged caution, which is precisely the opposite of what we need. Ambition at the centre is not enough, though. Competence is the *sine qua non*. This is about who chooses to go into politics, who is recruited by the civil service, how political decisions are made, and how agreed policies are then translated into measures that affect our lives. None of this has been working well.

As a starting point, it is hard to disagree with Starmer that moving power closer to people is a precondition of improving both the legitimacy and effectiveness of government. It means transferring more powers to the national governments of Scotland, Wales and Northern Ireland, and to English regions and mayors. As for the machinery of government, it is widely seen as a Victorian steam engine in an AI age. Mark Sedwill had the broadest and most important responsibilities of any recent civil servant, as Cabinet Secretary and National Security Advisor from 2017 to 2020. He warns that 'trying to transform the economy and society through an untransformed governance system is unlikely to prosper', that a system that would be 'familiar to Gladstone' could not possibly rise to today's challenges. He complains that central government is 'too metropolitan, too short-term, too siloed, too rivalrous and too focused on the preoccupations of Westminster and Whitehall.' He wrote the foreword to a report by the think tank Reform* that took

---

* 'Breaking Down the Barriers: Why Whitehall is so hard to reform', Reform, August 2023

evidence from twelve former permanent secretaries, six ex-cabinet ministers and many others who have struggled and too often failed to make a difference when working at the centre of government. Here is the exasperated voice of the erstwhile establishment complaining that the historic nexus between No. 10, the Treasury and the Cabinet Office on the one hand and the civil service on the other is bankrupt.

Technology will (or at least should) transform the wider public service, improving productivity, responsiveness, service levels and precision. Whitehall needs the capability to lead that transformation with sufficient diversity, including cognitive diversity, at the heart of government to understand the nation's varied communities, challenge perceived wisdom and innovate. This will require new accountability and incentive structures, more interchange with the private and third sectors, and more modular careers. A package along these lines would amount to the most ambitious peace-time reforms to Whitehall, the wider public service and the governance system since Attlee.

In his year and a half as a self-appointed destroyer of convention and complacency at the heart of government, Cummings put the reform of Whitehall as one of his more important ambitions, and failed, after his spectacular falling-out with Johnson. Now he believes that reconstruction can't happen unless and until a brand-new party buries the Conservatives after the next general election and takes the reins – he would hope – in 2028:

The Tory brand is horrific. The failure to do much other than cause chaos, their obvious lack of interest in productivity and growth, their appalling paralysis post-2020 on every important issue is much worse for them than they grasp. They've also trashed the reputation of 'capitalism'/free markets and strengthened the view that capitalism is a racket to help the privileged . . .*

He praises Sedwill sardonically for repeating a Whitehall critique redolent of one of his blogs from 2014, and makes a series of suggestions to restore confidence in the political system more widely. They include: a requirement to consider external candidates for all civil service and ministerial appointments, performance-related pay for MPs tied to average wages and living standards and instant publication of MPs' expenses and tax returns because he has 'a strong feeling there's a lot more corruption than is realised and many MPs think of themselves as justified in dodgy dealing because "my salary's too low".'

Whether you see Cummings as someone who moved the UK nearer to or further away from being bust, he is not to be dismissed lightly: he has been sounding the alarm for years, correctly, that our leaders have not been grasping the serious challenges to our way of life. Kish and I agree with his general point that 'more of the same' is the road to national collapse, while rejecting many of his prescriptions. However, our book is not a comprehensive and detailed manifesto to remake the United Kingdom. We've described some of what's gone wrong, and our intention is to encourage debate about policies that might bring recovery and rehabilitation. Kish and I aren't arrogant or stupid enough to say we have all or even some of the answers. But we are arrogant and stupid enough to urge you to open your mind to possible reforms that run counter to the consensus of the past half-century.

---

* Cummings on Substack, 11 August 2023

We're going to list some of those, and try to initiate a conversation about them on social media. But first I want to repeat that artificial intelligence, generative and narrower – and developments yet to come – changes almost everything. There are risks. But if it can be harnessed and rolled out rapidly, across the private sector, across the NHS, in schools, in universities, in the civil service, the benefits in respect of higher productivity and efficiency could be significantly enhancing to our prosperity and our health. It requires every company and every public service – at national and local levels – to set up taskforces to rapidly evaluate where AI can be adopted immediately, where roll-out requires structural institutional change, how human monitoring and know-how can be deployed vigilantly to stop things going wrong.

Every school student should be encouraged to use generative AI programs like ChatGPT inside the classroom, so that it becomes a tool to be deployed responsibly, rather than a homework cheat. And as and when there is a service that is plainly superior to the rest, if it's not free, the government should negotiate cheap access for every school student. As rapidly as possible, self-diagnostic AI services should be provided under an NHS brand, to massively accelerate evaluation of who needs to see a general practitioner and how rapidly treatment is required. And as the Tony Blair Institute says, we need to move much more rapidly to connect the NHS app on our phones to our patient records, which all need to be digitised. This information then needs to be aggregated and sorted to make it a precious resource for our life sciences industry. And an AI diagnosing program should have access to our personal data – our medical history, vaccination records, past and current medications – while it responds to whatever we ask it in a virtual consultation. As for businesses and public services that have direct contact with people, with clients, all of them need an AI 'friend' working with them, as soon as possible. When dealing with customers on the phone, the time saving and the reduced risk of error should both be valuable. What's vital though right from the outset is that AI is treated as servant and never boss. We need a culture in which AI is never gospel, never the ultimate arbiter, always the counsellor that is subject to verification by human intellect and human conscience.

The biggest priorities of all are those I discussed in chapter 3, namely redesigning the welfare and education systems for an AI industrial revolution. In summary, this is to change school education so that it is much more focussed on the non-robot interpersonal and creative skills, and to encourage resilience, adaptability and the ability to set up a business. For adults, our existing universities and further education institutions need reconfiguration to help people acquire the skills for brand-new careers, perhaps on multiple occasions, in a working life that may stretch fifty or sixty years. 'Lifelong learning' needs to become a high-quality, institutionally rooted service, rather than the empty catchphrase over every minister and their shadow. And the welfare system also has to be adapted to provide adequate financial support to those going through each one of these vocational changes. For what it's worth, I am much more excited than scared about artificial intelligence. If we are brave as a nation in how we adopt and control it, we can generate the resources that will fix our creaking social infrastructure.

We mustn't repeat the huge mistake of 1945, which was to be smug and complacent having won the War. As I said at the start of the book, the most important psychological change we can make is to accept defeat and then rebuild.

More than seventy-five years ago, our economic regeneration was not as bold and ambitious as that of France and Germany because we did not believe it had to be. We'd been victorious. Our system had triumphed. We had the legacy of Empire turning into Commonwealth. We thought ourselves special, still somehow ruling the waves, still almost a superpower, all objective evidence to the contrary. As a result, we weren't bold enough in our industrial reconstruction. We let Germany, in particular, overtake us in growth and dynamism too quickly. We would not tolerate the kind of sacrifices necessary to build a more formidable industrial base. And we were disappointingly short-term in our approach. There was very important societal transformation undertaken by the Attlee government, with the creation of the National Health Service, the extension of universal state education, the broadening and deepening of the welfare state and the benefits and pensions system. But there was no comparable rehabilitation of the wealth-creating part of the economy, the private sector and associated vital infrastructure.

Here, therefore, in no particular order, are a series of questions that any country facing our scale of challenge should be asking itself. In coming months, Kish and I will be asking for your views on them via various social-media platforms. We want to encourage an informed, generous-minded discussion, not the frequent, furious 'how can you be so thick!' social-media response. And to get our excuses in early, Kish and I don't have all the answers, but we believe we have a few relevant questions. Here they are.

GOVERNANCE

1) Our electoral system and uncodified constitution are letting us down. We need more long-term planning and cross-party collaboration to solve the economic and social dislocations that are imperilling our cohesion, our welfare, our prosperity and our happiness. We need to improve the calibre of our MPs and their willingness to work together, across party lines. So isn't this the moment to consider fundamental reforms of Parliament, including the introduction of proportional voting in general elections?

2) Should we consider cutting the number of MPs, increasing their pay but banning second jobs? We could reduce the number of MPs from 650 to 220. This would allow individual pay to increase from £90,000 a year to as much as £250,000 without additional burden on taxpayers. Money isn't everything in determining the quality of those putting themselves forward to be our elected representatives, but it is something when business leaders, charity bosses and head teachers can all earn significantly more than MPs. To eliminate the insinuation their views could be bought, *all* outside paid work would be prohibited. The reduction in their number would also help the parties and their members increase the proportion of candidates with talent, relevant qualifications and genuine commitment. They could be elected by strict proportional representation, though each would have formal links with people in an area that would be three times the size of current constituencies. They could be obliged, through arrangements like regular town hall meetings, to understand the needs of local people. But they would not do the constituency work, including surgeries, of current

MPs. They would concentrate more on policy work that applied to the whole UK, or England, in the case of devolved matters. See below for who would have the formal responsibility of representing individual voters.

3) Is it finally time to turn the House of Lords into a chamber of 650 elected representatives and no more than 150 experts, who would be appointed by a commission rather than political parties? The elected representatives could be chosen by the alternative vote system in individual constituencies, as currently configured. They would carry out the formal constituency representation duties that currently attach to MPs, and hold surgeries and be the main recipients of emails and letters from constituents. This upper chamber could be like the House of Lords in that it would be a chamber focussed on revising legislation that comes from the Commons, under the direction of the government. It would not be able to initiate legislation. And, just as today, there would be limits on its ability to veto the revealed will of the government.

4) Should the devolved powers of the assemblies in Wales and Northern Ireland be 'levelled up' to those of Scotland? All of them would have autonomous powers to alter the rates applicable to direct taxes for individuals and for businesses by plus or minus three percentage points. So, for example, the current corporation tax rate of 25% could be raised to as much as 28% in Scotland, Wales and Northern Ireland or cut to as low at 22%. However, the so-called Barnett formula of subventions from England to the nations would be calculated on the basis of the revenues that would flow from equalised UK tax rates. There'd be no corrective compensation if a nation reduced a tax rate and there was

a sharp fall in the revenues they received. The corollary would hold for a revenue increase from a tax rise.

5) Isn't there a case that there should be powerful mayors, modelled on today's metro mayors, though with powers extended to policing, transport, health and business investment, for the entire country?

6) Shouldn't there be pre-emptive protection for any risk that has catastrophic consequences, almost regardless of the assessed probability of the risk being crystallised? Isn't that the lesson of the global financial crisis, the Covid-19 pandemic and Putin's invasion of Ukraine? Surely we should employ the 'just-in-case' mindset, that is the justification for expensive standing armies and nuclear deterrents, when there is a risk of societal obliteration, such as from a new deadly disease or sabotage of the National Grid by an enemy like Putin or some unspeakable horror caused by a malevolent god-like AI? Arguably we should put in place substantial protections, regardless of whether the perceived probability is small. In that sense, the government's National Risk Register contains self-harming calibration, by providing an axis of spuriously precise probabilities for assorted threats. This axis of probability encourages officials and ministers, always wary of being accused of wastefulness, to reduce spending on any threat whose probability is small. The experience of the pandemic shows that there is no such thing as wasted pre-emptive expenditure on a threat of devastating magnitude.

7) Finally, in this section, an assertion, not a question. All government policies, including all of those in this chapter, should be subject to rigorous data analysis for their costs,

benefits, unforeseen consequences, and impact on inequality and prosperity, including important long-term costs such as climate change. All the data underlying policies could be published in both raw and interpreted form, to allow peer review by both experts and any citizen taking an interest. Rather than the black box it often is today, government would be an open-source project to promote the welfare of the UK.

SCHOOLS AND YOUNG PEOPLE

1) In the interests of encouraging a more intimate sense of the mutual dependence between citizens and government, and to help all young people understand the rights and responsibilities of being a citizen, shouldn't the prime minister and the leader of the opposition write a letter every January to everyone in full-time compulsory education, aged five to eighteen, explaining what they can expect from the state in their lifetimes and what the state expects of them? It would also be compulsory for these letters to be discussed in a special classroom session for at least an hour. It would not matter if the content of the letters was similar year to year.

2) Every child should surely be obliged to continue to develop their creative skills and abilities for their entire school careers, given that our creativity will continue to distinguish us from robots and artificial intelligence at least for the foreseeable future. Relevant studies and activities would include visual arts, music, drama, assorted crafts and cooking.

3) Should every child be obliged to bring a state-provided tablet or a privately owned smartphone into the classroom – the

opposite of what mostly happens today – and be encouraged to use a generative AI program while doing classwork and homework? This would be to normalise the use of AI, and turn it into a controlled tool, rather than a device, as at present, to game homework and the education system.

4) Should it be mandatory for every young person to have a bank account at fourteen and have access to limited credit from eighteen? These would be 'proper' bank accounts. The life skill of managing personal finances is underrated.

5) Should every child be endowed with a trust fund by the government at the age of ten, one that would be even more ambitious than the defunct ones created by Tony Blair and Gordon Brown? The fund would be the fulcrum for learning about how to invest and how to set up a business. Half the money could be invested via a publicly owned venture capital firm (see the economy section below), the rest in a stock-market indexed fund. None of the money could be withdrawn until the student was twenty-eight. From eighteen, investment decisions would switch from the fund managers employed by the state to the young beneficiaries themselves, to incentivise the acquisition of investment skills – though there would be a passive investment backstop for those who chose not to develop those skills.

6) We should surely mandate every school student aged fourteen to eighteen to spend a morning or afternoon each week for ten weeks helping older or disabled people in a social-care setting?

7) There is a strong argument for compulsory high-quality work experience for every school student aged fourteen to eighteen. This could take the form of three consecutive days

of placements per year with employers in either the private or public sector. The work experience would be supplied via organisations such as Speakers for Schools (full disclosure: I set up this education charity in 2009) so that young people from disadvantaged backgrounds have precisely the same opportunities as the children of well-connected parents.

HEALTH

1) For a maximum of five years, while backlogs are cleared, should everyone going private for a non-urgent treatment receive a tax credit for the cost of the treatment or the premium paid on health insurance? There would be an automatic sunset clause in legislation for this.

2) Perhaps any doctor or nurse trained in the UK who chooses to work abroad within fifteen years of qualification should be obliged to repay – as a twenty-year debt – the government subsidy in their training over and above the cost of a normal degree. There would be exceptions for work in poor countries and war zones, if formal permission was granted, or in other extenuating circumstances.

3) Should any doctor or nurse trained in the UK be obliged in a similar way to pay back the estimated cost to the public sector of their medical education if their private-sector earnings exceeded more than 25% of their NHS salaries?

4) There is a case for a significant pay rise for any consultant who commits to work during an entire career only for the NHS. The ceiling on their NHS pay could be as high as £300,000 a year. This is obviously less than the top paid

private doctors earn but should surely be enough to bind many more consultants to a life of exclusive public service.

5) There should be an accelerated programme to digitise all patient records and make them available through the NHS app. The information could also be used as training data for an NHS-branded and NHS-managed AI diagnostics service. All of us would have the option of using the AI diagnostics service as a possible pathway to an immediate consultation with a hospital specialist or to a pharmacist endowed with a new pharmaceutical prescribing power, without the need to go through a gatekeeper GP as at present.

6) Should digitised health data be aggregated, ordered and used as a powerful research tool for the UK's life sciences industry? The sheer size of the NHS should mean that this data would give UK pharmaceutical and medical businesses global competitive advantages.

7) Every time any of us sees a GP, we should be asked to provide a short online review of our experience for a digital service to be set up by the Department of Health. GP surgeries that consistently received low ratings would be contacted by a new quality-control agency and helped to improve. In extreme cases of underperformance, they would be placed under special measures and the management of the surgeries would be put for a remediation period in the hands of government-appointed specialists.

MONEY, TAX AND INVESTMENT

1) A quarter of a century ago, the UK had one of the most thriving pension-fund industries and stock markets in the world. They were mutually supportive and channelled billions of pounds a year to UK companies for productive investment. But well-meaning reforms, designed to protect pensioners from incompetent management and the kind of plundering carried out by Robert Maxwell, have had the unforeseen and unfortunate consequence of massively diminishing the flow of funds from UK pension savers to British companies. The reforms compelled pension funds to switch from holding shares to holding government debt. Today the UK stock market and individual pension funds are small and dysfunctional by international standards, and the UK economy has become too dependent on overseas investors. This has harmed UK productivity and growth. The reforms also created dangerous risks for the government's finances and the stability of the financial system by encouraging funds to engage in an unwise hedging strategy, known as Liability Driven Investment. These risks were exposed after Kwarteng's disastrous mini-budget when pension funds were forced into a fire sale of their holdings of UK government bonds. At the root of the problem is that UK pension funds are too small and too cautious. The government should therefore compel funds to merge and pool their risks, to create the kind of superfunds that have been so successful in Canada. It could be done by removing tax breaks for funds whose size is below a certain threshold. Such a consolidation to create institutions, each managing hundreds of billions of pounds, would achieve economies of scale and the institutional confidence to invest more in entrepreneurialism,

technology and infrastructure. Kish and I have been impressed by the diagnoses of the banker Michael Tory for the Tony Blair Institute.* But the big question is this: should taxpayers become joint underwriters of these superfunds along with employers? Should a creative government take on this potential liability, to re-energise our listless and lethargic economy?

2) The Bank of England should create a central bank digital currency, and encourage its take-up by banks and their customers. This would enable better oversight of monetary and economic conditions, in real time, and more sensitive and appropriate use of interest rates and other monetary levers (see page 161).

3) When inflation returns seemingly sustainably to 2%, there is a case for the Bank of England's inflation target to be changed from 2% plus or minus 1% to 2% plus 1%. The reform would allow interest rates to be a little lower than would otherwise be the case, to stimulate investment by businesses and growth.

4) The UK's fiscal rules, which are intended to constrain the amount any government can borrow, encourage too much short-termism and volatility in public service provision, especially investment in good quality schools, hospitals and transport. There is a case for a new framework that would provide the kind of medium-term stability needed for rational planning. There could be a re-worded version of the US system, by which MPs would vote to set the maximum size of the national debt in nominal terms, including a cushion

---

* Kakkad, Madsen and Tory, 'Investing in the Future: boosting savings and prosperity for the UK' Blair Institute for Global Change, May 2023

for shocks, for five years hence. Kish and I are aware that in the US presidential system this has periodically led to government paralysis and market volatility, when Congress has refused to authorise a new debt ceiling. But the risk of such dysfunction should be less in the UK's parliamentary system. A second reform would be to introduce a rule that the net value of the state – that is public sector assets minus debt – should grow on a three-year rolling schedule. The corollary would be that the state's net debt – Britain's current sorry predicament – should be reduced. This new target to increase the value of publicly-owned assets, as a counterweight to the longstanding obsession with gross debt, would help to dismantle the Treasury's toxic culture that favours privatisation and the minimum investment possible – a case put persuasively by the Resolution Foundation. There would be an important incentive to maximise investment in high quality projects.

5) Many richer people pay relatively low rates of tax because a large share of what they earn comes from an increase in the value of their assets or capital, and the rate of capital gains tax is lower than equivalent income tax rates. Shouldn't we therefore align capital and income taxes, so that the marginal tax rate for those who are richest is always higher than for poorer people and taxation is properly progressive?

6) Tax allowances for investment and research by companies should be reviewed by a commission of experts every five years to determine the rates of those allowances that would maximise the likelihood that the value of investment and research in the UK would be above the average for the G7 richest countries as a share of national income. The Chancellor would then set the allowances at those levels. This would be

similar to the way the Chancellor accepts the recommendation of the Low Pay Commission for the national minimum wage. Setting these tax allowances would always be the first priority when making decisions about taxation, because of the link to growth and productivity.

7) Could the government allocate £2bn a year for ten years for investment in digital technology and life sciences? It would be distributed by a publicly owned venture capital firm and through trusted VC partners.

8) Perhaps every takeover by overseas interests in the digital or life sciences sectors, and with a value of £50m or more, should be subject to scrutiny by a new 'national strategic interests' unit within the Competition and Markets Authority. The CMA's mandate could also be amended so that a new generalised assessment of 'national strategic commercial advantage' would be a factor in their more run-of-the-mill adjudications on mergers and acquisitions. This criterion should be of equal weight in their deliberations to competition. This would be separate and additional to the current requirement under the National Security and Investment Act to notify the government about acquisitions relating to seventeen sensitive areas, including artificial intelligence, energy and synthetic biology, among others.

9) Should there be a de facto windfall tax on the banks? This would take the form of ending payments to them of Bank Rate interest on two-thirds of the reserves they hold at the Bank of England. This would immediately yield a saving for the Bank of England, and therefore for the Treasury as underwriter of quantitative easing, of 5.25% on around £500bn. It would in effect raise more than £26bn a year for the government.

FAIRNESS IN THE ECONOMY

1) Should a set of maximum retail or shop and online prices for the necessities of life be mandated by a reborn government Price Commission? It would apply to a basket of food essentials, such as eggs, some vegetables and salad, bread, pasta, chicken and so on.

2) Every vital public service operating in an industry of natural monopolies – so water, energy, the postal service, rail and cable broadband – should surely revert to public ownership almost automatically, if they fail to meet reasonable agreed service and investment requirements over a rolling five-year period.

3) Should the threshold for when an employer must recognise the right of a trade union to negotiate on behalf of members be cut from 40% of all employees in a bargaining unit to 30%?

4) Whenever a person faces a compulsory change of career because of technological or profound economic shifts, shouldn't they qualify for a universal basic stipend for the months they would need to acquire important new skills? This stipend would be a high proportion of what they earned before needing to retrain.

5) Shouldn't universities and FE colleges be reconstructed to turn them into high-quality providers of sophisticated adult skills, for those needing a change of career? They would be as much about providing lifelong learning as pre-career education.

6) Surely everyone should have a digital identity card, necessary for access to all public services, which could store

important information, including health data and earnings? We know identity cards aren't thought to be British. And there are obviously security and civil liberties concerns. But in a time of scarce resources, it would help the government target support for those with greatest need in a time of crisis – such as the recent surge in energy prices – much more efficiently.

7) Should there be a three-month deadline for processing any application for asylum in the UK? After three months, if no decision had been taken, the applicant could be given a temporary identity card. This would allow the individual to work, but with a time limit on the card of six months. The identity card could be renewed on a rolling basis for as long as the asylum seeker's case was still under review, and subject to the individual not breaking the law. Work without a valid identity card, even part-time work, would be illegal. Asylum seekers allowed to work in this way would be paying tax and national insurance and would no longer be such a financial burden on the state.

8) A national chain of 'multibanks' should be created. Supermarkets and big retailers, including the likes of Amazon, would receive a modest tax credit in return for their donations of surplus food and other stock. Distribution of the stock to disadvantaged families and those on low incomes could largely take place, as per Gordon Brown's initial model, via social workers and the Citizens Advice Bureau.

9) On the twentieth anniversary of the vote to leave the European Union, in June 2036, when those born in 2016 have reached adulthood, shouldn't there be a referendum on whether to apply to rejoin the EU? This time there would

be a threshold suited to such enormous constitutional and economic change. It would, say, only have effect if 70% of the British people were to participate, and if a minimum two-thirds were to vote in favour. It would be far enough away not to undercut attempts to make a success of being outside the EU and would be near enough to enfranchise the majority of younger people who see Brexit as a mistake and a blight on their prospects.

Just to remind you, these are not the firm policies in the manifesto of the soon-to-be-created North London House party (hello Liz Truss). Some would be seen as of the left, some of the right. They are supposed to be provocative questions, to spark important conversations about how we lift the gloom and regenerate hope. Kish and I have tried not to repeat policies that are already pledged by the big parties, or already in mainstream conversation and debate. And you'll have read others of our suggestions in earlier chapters. This is not to say that we don't see the value of some or much of what many elected politicians are currently promoting – if not actually doing – such as the abolition of the two-child limit on universal credit payments or a significant extension of free school meals or a significant enhancement of the capability of mental health support, plus a gamut of measures to reduce emissions of greenhouse gases and improve the efficiency of energy use, or a whole host of tax reforms. It is simply that we want to challenge the imagination of those elected to govern. They need to make economic regeneration the priority above everything else, because it is the poorest and most vulnerable who suffer when a stagnating or shrinking economy deprives creaking public services of the vital funding they need.

Neither of us is gloomy or fatalistic. We have tried to show

how artificial intelligence, subject to diligent oversight, and bold investment in a new generation of industries can revive our prosperity. Also we have confidence that the United Kingdom will clear a new path for itself. It will be a path to a future in which we can feel pride, through supporting the neediest and most disadvantaged in our own communities and by promoting the values of mutual understanding and tolerance internationally. It is also important to remember that profound change does not require the kind of flamboyant leadership we witnessed in Boris Johnson. He's been replaced by a meticulous technocrat in Rishi Sunak, whose opponent across the Commons floor is another diligent technocrat. The political differences between them *seem* more to do with the different challenges they face in holding together the coalitions of views and interests that each of their parties represents, rather than anything fundamental about them - although one of Sunak's aides says the prime minister is 'much more ideologically right wing' than he has revealed to date. Each is devoted to their families. Sunak is by instinct a California liberal, Starmer a Camden Town liberal. The most conspicuous difference is in how they dress: West Coast sliders and ankle-length trousers versus a roomier and more sober North London suit. There is at the time of writing a significant argument between them on the speed and methods of moving the economy away from the production and use of hydrocarbons, on the optimal route to net zero. That is not a disagreement on the science of climate change, but is Sunak trying to bind in those Tory voters who don't want the short-term costs of changing their cars or central heating systems to those that emit less greenhouse gas, and don't see why the UK shouldn't continue to exploit oil and gas in the North Sea if other countries won't abandon their hydrocarbon reserves. Even though someone of Sunak's education and background could not possibly be a climate change denier, he is creating an almost ideological gap with Starmer over a global threat we're daily reminded of by horrifying pictures of fires and famine. But if the anger and passion stoked by their respective positions on climate action is anomalous, that doesn't mean neither are capable of pursuing courageous policies more generally, *after* the election.

It is worth remembering that expectations were low about what Clement Attlee could achieve in 1945 and Margaret Thatcher in 1979. Churchill denigrated Attlee as a 'modest man with much to be modest about'. Margaret Thatcher, speaking in 1973, said 'I don't think there will be a woman prime minister in my lifetime,' and there were plenty of members of the British establishment, on both sides of the political divide, who agreed with her, right up until the moment she became prime minister – and at that juncture they were convinced she would fail. I am not saying Starmer or Sunak will turn out to be Attlee or Thatcher. I am saying that millions of British people have had their hopes dashed that Brexit would put such an electric shock into the political and economic systems that their voices would at last be heard and their needs met. And when they then spun the wheel again by electing the most charismatic PM of modern times, Boris Johnson, they lost again. The lesson is that a revived Britain can't be achieved through populist catchphrases, or simple solutions to complex problems. What's required is rigorous analysis, strategic thinking, staying the course over the long term, and – Kish and I say this with all the self-denying sincerity we can muster – a recognition that tomorrow's headlines are of no significance whatsoever. It's the headlines in ten years that count.

ACKNOWLEDGEMENTS

ROBERT'S

I am one of those lucky people who feels excitement each morning at what my day has to offer. This is partly because – as loved ones routinely warn me – I am too addicted to work. It is also because ITV and *ITV News* are top-class institutions that produce the kind of high-quality news and current affairs that is the bedrock of any thriving democracy. I am proud to work for them. Particular thanks are due to Vicky Flind, Rachel Bradley, Michael Jermey and Andrew Dagnell, who mostly succeed in keeping me between what are fashionably called the 'guard rails' and stop me making too much of a fool of myself most days.

Biggest thanks are to my large, noisy, sometimes exasperating, always loving and supportive 'modern' family (in alphabetical order): Audrey, Charlotte, Douglas, Ed, Juliet, Margot, Max, Mum, Myra and Simon. They put up with me, somehow.

I don't really have appropriate words to express how grateful I am to Kish. He has been the perfect collaborator and co-author, as I knew he would be, simultaneously creative and meticulous. This is our book, though the stupid bits are obviously mine.

North London, 15 August 2023

KISHAN'S

First and foremost, my sincere thanks to Robert for this amazing opportunity. Your reporting on the 2008 crash made me want to study economics. The idea then I'd be working on *Bust?* with you now would have seemed truly surreal but I've learned so much doing it. To Rupert Lancaster, and all those involved in making this book a reality, thank you for trusting me to be part of it. To my teachers at every level who always offered encouragement and insight, I hope reading this is a boost rather than a worry that I never really did get it. To Vicky Flind, my brilliant TV mentor, you've taught me so much and reminded me not to take it too seriously all the time. I wouldn't be doing this or staying sane without you. To the joyous Peston team (past and present), Anushka Asthana and everyone at MultiStory Media and ITV who've very graciously tolerated my juggling of late, you make covering the madness of modern politics enjoyable. But most importantly my profound thanks must go to my nearest and dearest. To my parents who have always offered such grounding and inspiration, thank you for everything. To 'Chirm' and 'Aykel', I couldn't ask for better cheerleaders. And Lara, you're the strongest person I know. Everything I do is better because we're together. Yes, I will have more time for the pets now, honest!

South London, 15 August 2023

INDEX

Altman, Sam 59
Andrew, Prince 10
ARM chips 182–3
artificial intelligence
   biases in 53–5
   and chatbots 55–6
   and digital currencies 162–3
   economic impact of 37–42, 44–5
   and employment 37–42, 44–5
   in inflation forecasting 108–9
   language usage 33–6
   political response to 42–4, 48–9, 57–61
   potential harms from 45–51, 56–7, 59–62
   and productivity 36, 37, 39–40, 41, 51
   in public services 51–3
   in schools 249–50
   societal impact of 36–7
asset price rises 127–130
AstraZeneca 184–5
Atlee, Clement 268
Australian Broadcasting Corporation (ABC) 55

Bailey, Andrew 109, 154
Balls, Ed 154, 193–4
Bambra, Clare 83
Bank of England
   and Covid-19 outbreak 106–7
   forecasting model for inflation 108–9
   inflation target 152–6, 260
   and international cash flows 125–6
   and quantitative easing 128–9, 130, 131–4
   response to 2022/23 inflation rise 105–6
   sets interest rates 122–5, 133–4, 136–7
Bank Rate 123–5, 133–4
banking crisis (2008)
   effects of 14, 16
   preparedness for 101
   as turning point for economic progress 24–6
Belloc, Hilaire 4–5
Bernanke, Ben 109
Berner-Lee, Tim 192
Biden, Joe 40, 143
Big Bang 177–8
Bitcoin 159, 160
Blair, Tony
   life under 20–2, 25–6

and National Health Service
   85–7
and remaking of Labour Party
   23
Blanchard, Olivier 220
Bowie, David 18
Bradby, Tom 237
Brand, Paul 9
Brexit
   and Cambridge Analytica 50–1
   economic damage from 10, 16,
      27–8, 29
   and economic growth 26–7,
      29, 165–6
   and inequality 26–7
   and inflation 138, 149, 152
   as response to decline 14,
      25–6, 27–8
   and sense of exceptionalism
      242–3
   Vote Leave campaign 242
British Business Bank 192
British Patient Capital 192
Brooks, Rebekah 231
Brown, Gordon
   and Bank of England independence 122
   comparison with Boris Johnson
      239–240
   and disaster preparedness 100
   in Kirkcaldy 222–3, 238
   joining the Euro 196–7
   life under 25–6
Brown, Ophelia 185
Brynjolfsson, Erik 37
BT Group 45

Budd, Sir Alan 153
Buckland, Sir Robert 51
Burns, Terry 153
Busby, Siân 22
business start-ups 185–7

Cain, Lee 76, 231
Calver, Tom 216
Cambridge Analytica 50–1
Cameron, David
   and Covid-19 outbreak 80–1
   and inheritance tax 218
   relations with China 141
   sense of privilege 228–9
   straightforwardness of 229–30
Carnegie, Andrew 224
Carney, Mark 41–2, 212
Casey Review 15–16
Cash, Johnny 50
Cates, Miriam 207
Charles III, King
   arrests of volunteers after coronation 3, 8
   coronation of 6–8, 9
   as unifier 12–13
chatbots 55–6
ChatGPT 33, 37, 50, 250
Child Trust Funds 193–4
China
   increasing threat from 141–2
   manufacturing shift to 140–1
Churchill, Winston 268
Clark, Greg 122
COBR meetings 68
Coffey, Thérèse 88
Collaborations Pharma 56

Competition and Markets
    Authority 112, 263
Conran, Terence 178
constitution
  reforms of 253–6
  robustness of 22–3
Corbyn, Jeremy 14, 157
Cornell University 36
Cottage Family Centre 238–9
council tax 219
Covid-19 outbreak
  austerity effects on 79–81
  early responses to 66–8, 70–1,
    91–5
  and 'Eat Out to Help Out'
    71–3
  effects of 16
  Hallett inquiry into 70, 91,
    95–6
  health inequalities exposed by
    83–5
  and housing 211–12
  improvised responses to 87–8
  and inflation 137–8, 140–1,
    143–4
  initial reporting on 64–5
  interest rates during 106–7
  international comparisons
    69–70, 89–92
  and lockdowns 73–5
  long-term effects of 78–9
  media reporting on 89–90,
    96–7
  National Health Service during
    75–8, 79–80
  preparedness for 97–8

quantitative easing during
    130–3
  and SAGE advisory committee
    72, 73, 92–6
  and social care 81–2
  spread amongst MPs 88–9
  vaccination programme 68–9,
    240–1
Crick, Francis 192
Crisis 203
Crisis of Democratic Capitalism,
    The (Wolf) 12
cryptocurrencies 125, 158–63,
    261
culture wars 8, 11, 12
Cummings, Dominic 73–4, 89,
    145, 232–3, 240, 248–9

Dacre, Paul 231
Daily Mail 231, 235, 237
Daily Telegraph 230–2
Dear England (Graham) 242
decline, sense of
  and 1970s 18–20
  author's different experiences
    of 17–22, 23–7
  in Blair/Brown years 20–2
  and Brexit 14, 25–6
  and Charles III's coronation
    9–11
  and despondency 11–12
  and National Health Service
    79
  political effects of 14–15
  symptoms of 15–16
  under Margaret Thatcher 20

DeepMind 46–8, 182
Demos 218
Department for Science, Innovation and Technology 42–3
digital currencies 125, 158–63, 261
Dilnot, Sir Andrew 83
disaster preparedness 97–103, 255
Dogecoin 159
Donelan, Michelle 43–4
Dorries, Nadine 88–9
Doyle, Jack 232
Dury, Ian 18

'Eat Out to Help Out' scheme 71–3
economic growth
  and austerity 29
  and banking system 28–9
  and Brexit 26–7, 29, 165–6
  and business start-ups 185–7
  fall in 26–7
  and foreign competition 182–6
  impact of artificial intelligence on 38–9
  and investment 188–8
  Keir Starmer's policies for 166–7
  and living standards 169–72
  and migration 167–8
  research & development budgets 187–8, 191–2
  under Liz Truss 164–6
Edmunds, John 72, 73

education
  and productivity 192–3
  reforms to 256–8
Edwardes, Charlotte 64, 70, 89
Edwards, Albert 157
Elizabeth II, Queen
  death of 3–5
Eloundou, Tyna 38, 45
Engels' Pause 41–2
European Central Bank 111
Evans, Chris 230–1, 233–4

facial recognition software 53–4
Falcon-40B 61
Farage, Nigel 22, 240
Fetzer, Thiemo 72
Fife Big House 239
Financial Times 14–15, 59, 227
food prices 148–9
foreign competition 182–6
French, Simon 209
Friedman, Milton 146

Ganguli, Deep 53
GB Energy 157
Global Institute for Change 85
Goldman Sachs 37
Goodwin, Andrew 213
Gopinath, Gita 112
Gove, Michael 74–5, 201
government spending
  and interest rates 31–2, 118–22
  on National Health Service 79–80, 84–5
  as share of national income 31
GPT-4 33–6, 55–6

Graham, James 242
Grange, Pippa 242
Great Transfer 214–17
Green, Sir Owen 178
Green, Philip 176–7, 178
Grenfell Tower 200–1
Griffiths, Lord 131
Grimond, Jo 169–71, 173
Guardian, The 150
Guerin, Ben 76

Haldane, Andy 136, 190
Hallett inquiry 70, 91, 95–6
Halpern, Ralph 178
Hammond, Martin 226–7
Hancock, Matt 64, 65, 88, 90, 93, 96
Hanson Trust 178, 179–80
Hanvey, Neale 223
Harman, Harriet 227
Harri, Guto 235
Hassabis, Demis 58, 60, 182
Hattersley, Roy 145
Havers, Lord 180
Healey, Denis 156
Health Foundation 84
Hepburn, Audrey 179
Hewitt, Dan 200
Hogarth, Ian 59–61
Holland, Damian 88
homosexuality 19
honours system 2
Horizon (software) 52
housing
    during Covid-19 outbreak 211–12
and the euro 197–8
generational divide in 198–9, 206–8, 209–10
government strategies for 204–6, 214
and the Great Transfer 214–17
impact of rising prices 198–199
and interest rates 116–18
mortgage rate rises 135–6, 208–9
and planning system 203–4
and rental sector 200–1, 207–8
resilience against shocks 212–13
and social housing 201–3
taxation reform in 218–220
unaffordability of 199–200
How Do We Fix This Mess? (Peston) 28
Howe, Geoffrey 107, 146
Hsu, Steve 61
Hunt, Jeremy 80, 113, 148, 214, 246
hypothecated taxation 221

inequality
    and Brexit 26–7
    despondency over 11–12
    health inequalities 83–5
    and housing 214–17
    monarchy as symbol of 5–11, 12–13
    reforms to tackle 263–5
    and shoplifting 6–7
    and taxation 30
inflation

asset price rises 126–8
and borrowing and saving 115–16
and Brexit 138, 149
and company profits 111–13, 156–7
and Covid-19 outbreak 137–8, 140–1, 143–4
and debt 114–15
disinflation in 1990s 107–8
effects of 109–11
and food prices 148–9
and gearing 115
government policies to control 145–53
and international cash flows 125–6
and living standards 114–15
manufacturing shift to China 140–1
modelling for 108–9
as modern phenomenon 122
oil spike in 1970s 144–5
and quantitative easing 128–32
rise in 2022/23 104–5, 137
target setting 152–6, 261
and taxation 113–14
and Ukrainian invasion 139–40, 148–9
informal care 1–2
inheritance tax 217–18
Institute of Fiscal Studies 15, 114, 215
interest rates
 asset price rises 126–8
 and 2022/23 inflation rise 105–6
 Bank of England sets 122–5, 133–4, 136–7
 and Covid-19 outbreak 106–7
 and government spending 31–2, 118–22
 and house prices 116–18
 and international cash flows 125–6
 natural rate of 31–2
International Monetary Fund (IMF) 156
Iraq War 22
Irish Times 46
Ivens, Martin 234

Johns Hopkins University 69
Johnson, Boris
 and 2019 general election 14, 27
 and Brexit 16
 comparison with Gordon Brown 239–40
 and Covid-19 outbreak 67–8, 74, 75, 89, 94–5, 240–1
 encounters with author 227–8, 235–7
 on inflation 154
 and 'levelling up' agenda 27, 239
 Prime Ministerial style of 230–3, 240–2
 relationship with media 233–5
 sense of privilege 228–9
 and social care 83
 stands down as MP 225–7, 237–8

stands down as Prime Minister 4
use of honours system 2
Johnson, Carrie 231, 234
Johnson, Jo 227
Johnson, Paul 219
Johnson, Rachel 227
Johnson, Stanley 226

King, Mervyn 123, 153
King, Stephen 104
Kirkcaldy 222-4, 237–240
Krugman, Paul 130, 150
Kwarteng, Kwasi 121, 147, 164, 211

Lebedev, Evgeny 231
Legg, Shane 59, 182
Lehman Brothers 128
'levelling up' agenda 27, 239
Lewis, Martin 214
Liability Driven Investment 260
Libra 160
Lister, Ed 232
living standards
    decline since 2008 banking crisis 14
    despondency about 11
    and economic growth 169–72
    and inflation 114–15
    international comparisons 14–15
    and migration 168–9
    political barriers to 174–6
    political impacts of 14
    and productivity 169–72
    in public sector 15

Markus, Billy
Marley, Bob 18
Marmot, Michael 83, 84
May, Theresa 240
McKinsey 39, 40
Metropolitan Police 15–16
Milne, Seamus 89
Modern Monetary Theory 129–30
monarchy
    as global advert 9–11
    as symbol of inequality 5–11, 12–13
monetarism 146–7
Monetary Policy Committee 108, 109, 122, 154
Monty Python 18
Moore, Sir Richard 57, 58
Morgan, Emily 88
Mostaque, Emad 55
Murdoch, Rupert 20, 229, 234
Murphy, Ken 7

Nandy, Lisa 202
National Health Service
    during Covid-19 outbreak 75–8, 79–80
    and health inequalities 83–5
    new technologies in 85–7
    public attitudes to 78, 79
    reforms to 258–60
    spending on 79–80, 84–5
    staff morale in 77
    waiting lists in 16, 77–8
National Housing Federation 203
National Risk Register 99, 100–1, 102

New Yorker 149, 150–1
Nightingale Hospitals 76
Nuffield Foundation 188–9, 215
Nuffield Trust 81
Nurse, Sir Paul 187, 192
Nvidia 62

Office for Budget Responsibility (OBR) 29, 31, 109, 119–20, 121, 166
Office for National Statistics (ONS) 69, 84, 118, 171, 187, 205, 206, 207, 239
Open AI 33, 59
Orwell, George 200
Osborne, Sir Cyril 169–70, 173
Osborne, George
   and Covid-19 outbreak 80
   effect of austerity 15, 29
   encounters with author 228
   government borrowing under 118
   relations with China 141
   sense of privilege 228–9
Oxford Economics 213

Palmer, Jackson 159
Pandemic Diaries (Hancock) 66
Patel, Priti 217
Perkins, Dario 165
Peston, Juliet 3, 6
Peston, Maurice 145
Philp, Chris 54
policing 15–16
Post Office 52
productivity
   and artificial intelligence 36, 37, 39–40, 41, 51
   and banking system 28–9
   business practices as barrier to 176–7, 178–82
   decline in 26
   and educational reform 193–4
   international comparisons 171–2
   and investment 188–8
   and living standards 169–72
   and migration 167–8
   political barriers to 174–6
   and private equity companies 181–2
   research & development budgets 187–8, 191–2
   and taxation 30, 31–2
   trends in 172–3
   uneven distribution of 189–91
Public Health England 93
public sector
   artificial intelligence in 51–3
   in government spending 121–2
   wage stagnation in 15, 158
Putin, Vladimir 98, 99

quantitative easing 128–34

racism
   in 1970s 18
   authors' experience of 3, 18
   and facial recognition software 53
Rees-Mogg, Jacob 173, 194, 217
Reform 247–8
Replika 55

research & development budgets
    187–8, 191–2
Resolution Foundation 26, 168–9,
    188–9, 219
Ricardo, David 143
Road to Wigan Pier, The (Orwell)
    200
Roberts, Simon 7
Royal College of Emergency
    Medicine 77

SAGE advisory committee 72, 73,
    92–6
schools
    absences from 16
    artificial intelligence in 249–50
    reforms to 255–7
    Schroders 198
Sedwill, Sir Mark 96, 247–8, 249
Sex Pistols 18
shoplifting 6–7
Slack, James 230, 231, 232
Smith, Adam 221–2, 244
social care 81–3
Social Democratic Party (SDP) 23
social housing 201–3
Softbank 183
Southgate, Gareth 242
Spectator, The 228
stamp duty 206, 211, 219
Starmer, Keir
    on artificial intelligence 42
    on Charles III's coronation
        7–8, 9
    economic growth policies
        166–7
    and inflation targets 156
    moves away from nationalisation 157
    technocratic style 266–8
Suleyman, Mustafa 182
Sun, The 231
Sunak, Rishi
    and artificial intelligence 42,
        58
    and Brexit 16
    on business start-ups 185
    on Charles III's coronation 7, 9
    during Covid-19 outbreak 71,
        73, 210–11
    government borrowing under
        118–19
    and housing 210–11
    and inflation 113, 121, 154
    relations with China 142
    use of quantitative easing
        130–3
    technocratic style 266–8
Sunday Times, The 64

Taiwan 98, 99
taxation
    on bank profits 134, 263
    hypothecated taxation 221
    and inequality 30
    and housing 218–220
    inheritance tax 217–18
    and inflation 113–14
    international comparisons 30–1
    reforms to 259–62
Taylor, Matthew 77
Thatcher, Margaret

effects of policies 20, 23–4
and inflation 107, 108, 146
taxes bank profits 134
Times, The 54, 68, 74–5
Tirole, Jean 220–1
Tony Blair Institute 250, 261
Tory, Michael 261
Trump, Donald 241
Truss, Liz
  becomes Prime Minister 4
  and budget of 2022 10, 121, 164, 165–6, 210, 245–6
  and economic growth 164–6
  inflation controls 130–1, 147–8
  and productivity 173
Tucker, Emma 64
Turing, Alan 192
Tweneboa, Kwajo 200

Ukrainian invasion
  effects of 16
  and inflation 139–40, 148–9
  preparedness for 97–8
  use of artificial intelligence in 50
Urbina, Fabio 56

Vallance, Sir Patrick 48–9, 91, 94
Van-Tam, Jonathan 90

Warner, Ben 73
Watson, James 192
Weber, Isabella 149–51
Wheeler, Charles 228
Wheeler, Marina 228
White, Gordon 179
Whitty, Chris 66, 91, 94
William, Prince 202
Williams, Robert 53
Wilson, Harold 170–1
Wittgenstein, Ludwig 34
Wolf, Martin 13
workforce absences 16
World Economic Forum 228–9
World Health Organization (WHO) 67
WTF? (Peston) 25, 28, 194, 196, 200

Xiao Zhi 3.0 51

Zahawi, Nadhim 217
Zelenskyy, Volodymyr 50